NEAR-INFRARED ORGANIC MATERIALS AND EMERGING APPLICATIONS

NEAR-INFRARED ORGANIC MATERIALS AND EMERGING APPLICATIONS

ZHI YUAN WANG

CRC Press
Taylor & Francis Group
Boca Raton London New York

CRC Press is an imprint of the
Taylor & Francis Group, an **informa** business

CRC Press
Taylor & Francis Group
6000 Broken Sound Parkway NW, Suite 300
Boca Raton, FL 33487-2742

First issued in paperback 2019

© 2013 by Taylor & Francis Group, LLC
CRC Press is an imprint of Taylor & Francis Group, an Informa business

No claim to original U.S. Government works

ISBN-13: 978-1-4398-6193-6 (hbk)
ISBN-13: 978-0-367-38007-6 (pbk)

Library of Congress Cataloging-in-Publication Data

Wang, Zhi Yuan.
　　Near-infrared organic materials and emerging applications / author, Zhi Yuan Wang.
　　pages cm
　　Includes bibliographical references and index.
　　ISBN 978-1-4398-6193-6 (hardback : acid-free paper)
　　1. Infrared technology--Materials. 2. Organic compounds--Spectra. 3. Near infrared spectroscopy. I. Title.

TA1570.W36 2013
621.36'2--dc23   2013008017

Visit the Taylor & Francis Web site at
http://www.taylorandfrancis.com

and the CRC Press Web site at
http://www.crcpress.com

Contents

Preface

> The essence of knowledge is, having it, to apply it;
> not having it, to confess your ignorance.
>
> **Confucius (551–479 BC)**

This book is in essence about the knowledge and applications of invisibly "colored" organic materials. It may seem to originate from my childhood fascination with colors and color changes, which later motivated me to acquire knowledge in chemistry and science throughout my high school and university studies. After nearly two decades of learning, studying, and playing with many "colorful" compounds and polymers, I came to realize that there is a whole new range of fascinating, invisibly "colored" materials for us to explore and utilize. This life is wonderful and colorful; either you see it or not.

Given that the subject of near-infrared absorbing dyes has been documented in a number of comprehensive reviews and books since the 1990s, my approach to writing this book intends to focus on the relationship between optical properties and structures of this class of scientifically interesting and yet practically useful materials, as well as the molecular design principles and some emerging applications. I hope that my approach is helpful for readers, namely students, teachers, and researchers interested in the potential interplay of various aspects of visually colored and "invisibly colored" near-infrared organic materials and emerging applications. It is, therefore, written as a book to be read for the purpose of stimulating ideas and generating scientific curiosity, as well as a reference work to be used in research and development.

In this book, the major emphasis is on the chemistry of materials, in particular the structure–property relationship of organic compounds and polymers. Materials are universal and not outdated rapidly, while new materials are being discovered, introduced, and utilized all the time. Following a brief introduction (Chapter 1), the molecular design aspect is discussed in Chapter 2. Chapters 3 and 4 cover up-to-date information on the structures and key properties of near-infrared organic compounds and polymers, with the intention of providing the readers with both useful data and some interesting food for thought. The final chapter includes some examples of emerging applications of near-infrared organic materials. As Confucius taught us a long time ago, "The essence of knowledge is, having it, to apply it." I am in favor of both curiosity-driven and application-oriented study and research.

There are many people I want to thank. I must begin my thanks with the graduate students at Carleton University in Canada and those at the Changchun Institute of Applied Chemistry (CIAC) in China, who inspired this work through their devotion to scientific research and enthusiasm for learning in the past two decades.

My wholehearted thanks go to my colleagues and friends, in particular Professors Xinhua Wan (Peking University), Lixiang Wang, Dongge Ma and Donghang Yan (CIAC), for their encouragement, useful discussions, and friendships. These academic and personal interactions helped initiate my book writing and shape the way I put the book together.

I am indebted to my Ph.D. supervisor, late Professor George Just, and my postdoctoral research supervisor, Professor Allan Hay. Their dedication to education and research, philosophic views on science and life, curiosity, and need, thirst, or desire for knowledge taught me how to learn and to live.

I am very grateful to my wife Jane (Jian Ping Gao) and my sons Adam and Kevin. They endured the many days and nights of my being in the office and the endless hours on the computer with patience and understanding. My family constantly provides the love and the help required for me to keep learning out of curiosity and motivates me to apply knowledge in life. I dedicate this book to my family.

Zhi Yuan Wang
Department of Chemistry
Carleton University
1125 Colonel By Drive
Ottawa, Ontario K1S 5B6
Canada

About the author

Zhi Yuan (Wayne) Wang was born in Beijing in 1957 and moved around in his childhood to several cities in China to attend high school. He received his B.S. in chemistry from Peking University in 1982. He then entered graduate school in Toronto at McGill University, where he received a Ph.D. in organic chemistry under the auspices of Professor George Just. After working as a postdoctoral fellow with Professor Allan S. Hay at McGill University on high-performance polymers, he moved to the Canadian west coast and joined Ballard Power Systems in Vancouver. Shortly after one year, he went back for more postdoctoral training in Professor Hay's group. In 1991, he joined the faculty of the Department of Chemistry of Carleton University, where he is currently professor of chemistry and holds a Canada Research Chair in Emerging Organic Materials since 2001. He was elected to a Fellow of the Chemical Institute of Canada in 2003.

Professor Wang's research interests focus on the study and development of fundamentally important and practically useful organic materials, including nonlinear optical chromophores and polymers, near-infrared chromogenic materials, chiroptical materials, and photocurable polymers.

chapter one

Introduction

Colours speak all languages.

Joseph Addison (1672–1719)

1.1   Color and near-infrared light

Color is a highly multifaceted phenomenon in nature, science, and culture. Even the term *color* itself is difficult to be precisely defined. A prominent English essayist, Joseph Addison, once wrote "Colours speak all languages," while in a recent edition of *Webster's Encyclopedic Unabridged Dictionary* (1994), there are 24 different meanings for the noun "color" and 5 for the verb. To physicists and chemists, color means light, emission, absorption, spectrum, coloration, etc. Yet, "color" is also used widely in the context of many phenomena primarily for perceptual effects of other senses in nonvisual cultural activities such as music, poetry, and fiction, and as well as business and political activities. Primary colors is a well-established term in color science and in art, and simply means a set of colors, normally red, yellow, and blue for subtractive mixing and red, green, and blue for additive mixing, from which all other colors may be obtained by mixing. By coincidence, two books with almost identical titles associated with primary colors were in New York within 18 months of one another. *The Primary Colors* by Alexander Theroux (Henry Holt, New York, 1994) does consist of three essays on culture and other aspects of blue, yellow, and red. The book of *Primary Colors* (January, 1996) by "Anonymous" later revealed as writer Joe Klein is a *roman à clef*, a work of fiction that purports to describe real-life characters and events—namely, Bill Clinton's first presidential campaign in 1992.

Human endeavor with color can be dated back to more than 37,000 years ago to France's Chauvet cave paintings (pictures) or even more than 40,800 years ago to El Castillo cave paintings along northern Spain's Cantabrian Sea coast. After a long period of thinking of colors as mixtures of black and white based on the hypothesis of Aristotle (384–322 BC), the foundations of modern color research were laid by Isaac Newton (1642–1727). In 1666, he found that an inverted prism positioned after the first would recombine these colors into achromatic light and a second, noninverted prism was not able to split any of the colored components obtained after the first prism any further. From his experiments, Newton recognized the relationship between light and color, and also color's non-objectivity. However, nowadays we still experience colors through an extremely complex path of physical, chemical, neurological, and mental processes. We prefer to say (or even believe) "the blue sea," "blue sky," "green leaves," and "white wine," as we still convey the impression that color is a property that these things really possess. As shown on the front inside page, the near-infrared (NIR) photograph "Perfect Snowy Trees" may give one a surreal, ethereal feeling by seeing or believingly seeing the same true colors of the foliages and snow beyond the human visible vision.

The 19th century saw the most important step toward understanding the nature of light. James Clerk Maxwell predicted that light is a combination of magnetic and electrical phenomena. In 1873, after many years of working with models and developing theories, he implicitly abandoned yay or nay the necessities of using mechanical models and concluded that light and electromagnetism must ultimately be the same in nature, and so must both be waves of electromagnetic radiation, based on famous Maxwell's equations. A decade after Maxwell, Heinrich Hertz (1857–1894) conducted experiments on electromagnetic wave propagation in air and discovered short-wavelength radio waves were of the same nature as light. His finding enlarged the electromagnetic wave spectrum enormously (Figure 1.1). As shown in Figure 1.1, the visible or color spectrum covers only a minute portion of the entire electromagnetic waves known today and is adjacent to the ultraviolet and infrared spectral regions.

Infrared radiation is an electromagnetic radiation whose wavelength is longer than the longest wavelength of red color in visible light, but shorter than that of microwave. In 1800, William Herschel (1738–1822) used a prism to break up white light into different colors and then used a thermometer to measure the temperature of each of the different colors. He found that each color did not have the same temperature, and a thermometer placed just beyond the red end of the visible spectrum gave a higher temperature reading than the visible spectrum. Further experimentation led to Herschel's conclusion that there must be an invisible form of light beyond the visible spectrum and he named this invisible color as *infrared* (the prefix from the Latin *infra* means "below").

Herschel's work also implies a relationship between the property of light and temperature. Indeed, there is an inverse relationship between the wavelength of the peak of the emission of a blackbody and its temperature when expressed as a function of wavelength, which is often called *Wien's displacement law*. This law is expressed by an equation of $\lambda_{max}T = b$, where λ_{max} is the peak wavelength, T is the absolute temperature of the blackbody, and b is a constant of proportionality, equal to 2.8977685×10^{-3} m K or 2.8977685×10^{6} nm K. From this equation, it can be predicted

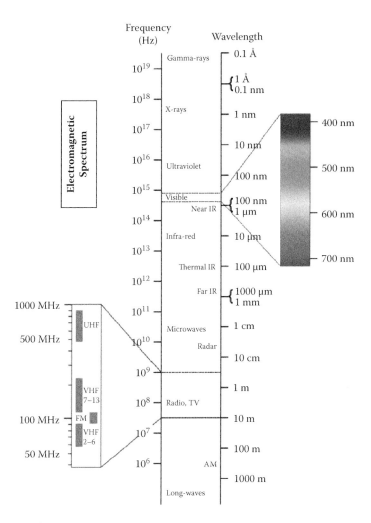

Figure 1.1 Electromagnetic spectrum.

that the hotter an object is, the shorter the wavelength at which it will emit most of its radiation. Since the effective temperature of the sun is 5778 K, the peak emission of the light form the sun is calculated to be at the wavelength 502 nm using Wien's law. This is the wavelength of green light. A wood fire at 1500 K puts out a peak radiation at 2000 nm, which is far more energy in the NIR region than in the visible region. The temperature of the living human body and mammals is roughly at 300 K, and most mammals emit radiation in the far-infrared region at 10 µm, which falls in the range of infrared wavelengths that pit viper snakes and HgCdTe-based IR sensors can detect.

Within the infrared spectral region in the electromagnetic spectrum, the infrared band is often subdivided into smaller sections for certain applications. The International Commission on Illumination (CIE) recommended the division of infrared radiation into the three bands: IR-A (0.7–1.4 µm), IR-B (1.4–3 µm), and IR-C (3–1000 µm). ISO 20473 scheme divides the infrared radiation into near-infrared (NIR) (0.78–3 µm), mid-infrared (MIR) (3–50 µm) and far-infrared (FIR) (50–1000 µm).

There are also several application-based division themes, such as astronomy division, and sensor response division [1]. For astronomers, the three regions are typically used for observation of different temperature range or different environments in space and the division boundary is not precisely defined, namely, (0.7–1) to 5 µm for NIR, 5 to (25–40) µm for MIR, and (25–40) to (200–350) µm for FIR. According to the response of the materials used in sensors or detectors, the infrared spectrum is divided into more than three regions, typically 0.7–1.0 µm as NIR, 1.0–3 µm as short-wave infrared (SWIR), 3–5 µm as mid-wave infrared (MWIR), 8–12 or 7–14 µm for long-wave infrared (LWIR), and 12–30 µm for very-long wave infrared (VLWIR). In most commercial products, silicon is used in a NIR sensor, InGaAs in a SWIR sensor, and HgCdTe in MWIR and LWIR sensors. In optical communications, the part of the infrared spectrum is divided into seven bands based on availability of light sources transmitting/absorbing materials (fibers) and detectors [2]. The C-band is the dominant band for long-distance telecommunication networks. The S and L bands are based on less well-established technology, and are not as widely deployed.

Band	Description	Wavelength range (nm)
O band	Original	1260–1360
E band	Extended	1360–1460
S band	Short wavelength	1460–1530
C band	Conventional	1530–1565
L band	Long wavelength	1565–1625
U band	Ultra-long wavelength	1625–1675

The boundary between visible and infrared light is not precisely defined. The human eye is markedly less sensitive to light above 700 nm, thus longer wavelengths make insignificant contributions to scenes illuminated by common light sources. But particularly intense light (e.g., from lasers or from bright daylight with the visible light removed by color filters) can be detected up to approximately 780 nm and will be perceived as red light. The onset of infrared light is defined, according to different standards, typically between 700 nm and 800 nm. In the fields of chemistry and biology nowadays, the NIR spectral region is typically recognized to be from 750 nm to 2500 nm.

1.2 Types of NIR materials

NIR materials are defined as the substances that interact with NIR light, namely, absorption and reflection, and emit NIR light, under external stimulation such as photoexcitation, electric field, and chemical reaction. NIR materials can be divided into two groups: inorganic materials including metal oxides and semiconductor nanocrystals (NCs) and organic materials such as metal complexes, ionic dyes, π-conjugated chromophores, and charge-transfer chromophores.

Semiconductor NCs are an emerging class of NIR materials and can be readily produced by colloidal synthetic methods in an aqueous or organic medium. It is highly feasible to tailor the optical and electronic properties of NCs for some specific applications by altering their chemical compositions and sizes. There is a strong dependence of the energy gap (E_g) between conduction and valence levels when the particle diameter approaches the scale of the de Broglie wavelength of an excited electron within the particle. The size-dependent band gap energy of semiconductors can be calculated, typically for some of II–VI, III–V, and IV–VI semiconductors with particle sizes ranging from 3 nm to 10 nm [3]. Semiconductors NCs with photon energies corresponding to the telecommunications wavelengths of 1.3 and 1.55 μm are particularly of interest and potentially useful for optical amplification (e.g., fiber amplifier) and NIR signal detection or night-vision view. Accordingly, HgSe, HgTe, InAs, and PbX (X = S, Se, Te) NCs are intensively and extensively explored for NIR applications.

Many transition metal-oxide films are electrochemically active and thus express electrochromism [4]. Usually, the reduced forms become highly colored and also exhibit intense absorption in the NIR region [5]. Among many metal oxides, such as WO_3, MoO_3, TiO_2, Ta_2O_5, and Nb_2O_5, that show color with electron injection and charge-balancing ion insertion, WO_3 is well studied and the most employed one for energy-saving smart windows. Its color can switch from intense blue in the reduced form to colorless or slight blue in the bleached state. The reversibility of color switch depends on the ion insertion coefficient, which is usually less than 0.2, or the applied voltage (Figure 1.2) [6]. It is obvious that the transmittance can be varied gradually and reversibly in the NIR region by the use of voltages from 0 to 2 V. The tungsten oxide film absorbs intensely above 1000 nm at the applied voltage of −0.8 V or higher.

Typical NIR-absorbing organic materials contain extended π-conjugated chromophores and charge-transfer chromophores, such as polyenes, polymethines, azo dyes, cyanines and merocyanines, donor-acceptor aromatic

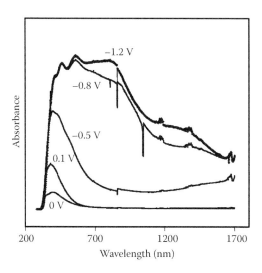

Figure 1.2 Spectral absorption as a function of applied voltage for the WO_3 film sputtered on ITO-coated glass in contact with electrolyte solution.

compounds, charged and radical quinones, and other compounds [7]. Metal complexes are another important class of NIR-absorbing organic materials, as represented by phthalocyanines, naphthalocyanines [8], mixed-valence dinuclear metal complexes [9], and nickel dithiolenes [10].

The dithiolene complexes of transition metals with the general structure shown in Figure 1.3 are the first reported NIR-absorbing metal complexes and show intense and broad NIR absorption in their neutral and anion states. Since the chemistry of metal dithiolenes and various related topics have been reviewed and well documented, this book does not intend to describe the same subject in-depth. Metal dithiolenes have aroused interest for some reasons: (a) their intense absorption in the NIR region, (b) their redox property, (c) high thermal and photochemical stability, and (d) capability of tuning the absorption wavelength by introducing substituents to the ligands. One of characteristics of dithiolene complexes, distinguishable from organic NIR dyes, is their broadband absorption without fine vibrational structure [11].

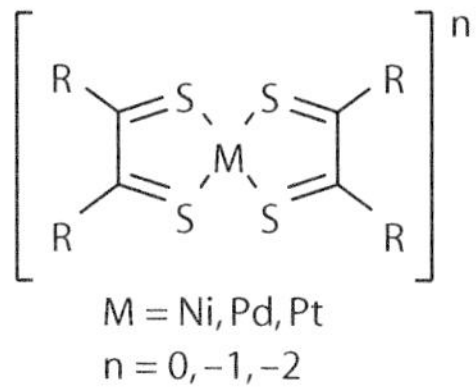

Figure 1.3 General structure of metal dithiolenes.

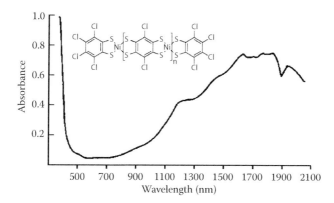

Figure 1.4 Absorption spectrum of Ni dithiolene coordination polymer. (From J. Campbell et al., *Dyes Pigm.* **1991**, *15*, 15. Reproduced with permission.)

The strong NIR absorption of Ni dithiolene complexes was assigned to π-π transition by Schrauzer and Mayweg [12] or to the L(π)-π* transition by Gray et al. [13]. More recently, Weighardt et al. argued that the Ni(II) remained in all species but the ligands involved electron transfer giving rise to the dianion and radical anion [14]. Numerous dithiolene derivatives were synthesized and used to form the complexes with Ni, Pd, and Pt. The NIR absorption ranges from 700 to 1400 nm with molar extinction coefficients up to 80,000 $M^{-1}cm^{-1}$, depending on the nature of ligands. By extension of the conjugation and coplanarity of metal–ligand centers and bridging ligands, the interaction between orbitals of metal and ligand can be greatly enhanced, resulting in a significant reduction of the energy gap and redshift in absorption into the NIR region. As shown in Figure 1.4, the Ni dithiolene polymer exhibits a low-energy electronic transition (from 900 to 2100 nm) with an extinction coefficient up to 2,000,000 dm^3 mol^{-1} cm^{-1} [15]. The very low energy of the electronic transition in this polymeric chromophoric system is lower than any previously reported metal dithiolene complexes, which brings an intriguing question as to how far this effect can be pushed into the mid-infrared and what will be the electronic and redox properties of such substances.

1.3   Current uses of and needs for NIR materials

The advance in infrared technology such as infrared imaging is driven by the needs in military and civilian uses. Military applications include target acquisition, surveillance, night vision, homing, and tracking. Nonmilitary uses include thermal efficiency analysis, remote temperature sensing, short-ranged wireless communication, spectroscopy, and weather forecasting. Research on NIR materials and technology is motivated by curiosity in fundamental study and practical applications in a number of important sectors such as energy, communication, bioimaging, sensing, and advanced optoelectronics. Nearly 50% of solar energy falls in the NIR spectral region, which calls for research on NIR absorbing dyes and photovoltaic materials in order to achieve a higher efficiency for solar energy conversion. As the NIR light penetrates deeper into the tissues, it is clearly advantageous to use NIR materials as tags or stain for bio-imaging. Using the NIR light for sensing and detection offers additional benefits for being invisible and less interference.

NIR-absorbing organic materials account for the most known NIR materials, and their applications are diverse and extensive in many technological sectors. For example, NIR-absorbing pigments are used in laser printers and digital copy machines as the charge-generation material [16], and in optical disks (e.g., CD-R) as the information-storage material [17]. NIR-absorbing materials can simply be used as a heat absorber to block the heat or as an optical filter to shield the NIR rays from electronic appliances. For example, plasma display panel television (PDP TV) radiates NIR rays of 850–1100 nm that can cause malfunctions of the remote control devices. Accordingly, the nickel dithiolene complex that absorbs strongly in the same NIR region is applied on the front surface of PDP TV to reduce the transmittance of NIR radiation [18]. In the medical applications, NIR dyes have been used as photosensitizers for photodynamic therapy in the treatment of cancer [19]. At present, one of the key limiting factors for organic solar cells is the mismatch of the absorption spectrum of the active layer and the solar emission spectrum. The use of NIR-absorbing or NIR photovoltaic small molecules and polymers could extend the materials' absorption into the NIR spectral region and even beyond a 1000-nm wavelength, which in theory could double the current power conversion efficiency of organic solar cells.

NIR emissive organic materials are relatively less available than NIR-absorbing organic materials, especially for those that emit the light at the wavelengths beyond 800 nm. This emerging class of organic materials is receiving more attention for their potential applications in telecommunication, display, and bio-imaging. Current optical communication networks use the NIR light at 1310 and 1550 nm for signal transmission and processing [20]. Making a device based on organic materials that emits efficiently the light at the telecommunication wavelengths is a great challenge. In addition, NIR emissive organic materials can offer some advantages in biological optical imaging, because of no or low NIR autofluorescence from tissue and deeper penetration of light into the tissues at wavelengths between 650 and 900 nm [21]. Moreover, NIR emissive organic materials may find applications in night-vision target identification [22], information security display, and sensors [23].

Studying how objects reflect the NIR light is just as important as studying an object's emission, transmission, and absorption of NIR light (the Sun's near-infrared radiation) for many applications. Typical applications using NIR refection spectroscopy include pharmaceutical, medical diagnostics, food and agrochemical inspection or quality control, observation of soil composition, as well as research in combustion, functional neuroimaging, sports medicine and science, urology, and neurology.

NIR reflective materials are widely used as coatings for an IR reflecting mirror or a heat reflector for low-E glass in buildings [24]. NIR reflective organic materials can in principle be easily processed for a variety of energy-saving applications, such as coatings on the decks of ships, clothing, tents, refrigerators, and other indoor and outdoor goods to reflect solar energy, minimize solar-induced heat build-up, and thus to reduce the energy consumption for cooling. However, the majority of NIR reflective materials are metals (e.g., silver and gold), metal oxides (e.g., TiO_2), and inorganic pigments [25]. NIR reflective organic materials are rare. Two patents claim the NIR reflective pigments containing copper phthalocyanine [26] and azo pigments [27]. When incorporated with commercially available pearlescent mica, the reflectance of halogenated copper phthalocyanine pigment can be over 50% in the wavelength range of 1000 to 1500 nm and maintains above 35% from 1500 to 2200 nm. However, the role that mica material may play or a possible band gap structure formed between mica and pigment are unclear. The azo compounds depicted in Figure 1.5 can be used as textile printing agents, paints and printing inks with a dark green color matching well to the green leafs, implying a possible use as a camouflage material.

It is well known that a green leaf can reflect both green and NIR light from sunlight, and the unique reflective spectral characteristics of different leaves can be used to identify the types of plants (Figure 1.6). Healthy

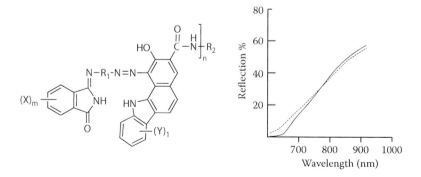

Figure 1.5 Structure and IR reflectance of azo compounds. Broken line: green plant; Solid line: azo compounds.

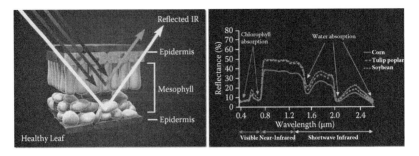

Figure 1.6 (See color insert.) Reflection of a green leaf, chlorophyll, and other plants. (From http://missionscience.nasa.gov/ems/08_nearinfraredwaves.html.)

vegetation absorbs blue- and red-light energy to fuel photosynthesis and create chlorophyll. The presence of more chlorophyll in a plant means more reflected NIR energy or a healthy and productive plant. However, although one may attempt to attribute such an ability of NIR reflection to the active ingredient chlorophyll in a green leaf, the reflection mechanism is still unknown and a possible supramolecular structure in a leaf that may play a role in NIR reflection is not yet confirmed [28]. The unique patterns of NIR reflection spectra maybe used as spectral signatures for vegetation to identify types of plants. For example, the reflection spectra in Figure 1.6 show the differences among the spectral signatures of corn, soybeans, and Tulip Poplar trees.

References

1. J. L. Miller, *Principles of Infrared Technology: A Practical Guide to the State of the Art*, New York: Van Nostrand Reinhold, 1994.
2. R. Ramaswami, Optical Fiber Communication: From Transmission to Networking, *IEEE Communications Magazine*, pp. 138–147 (May 2002).
3. M. T. Harrison, S. V. Kershaw, M. G. Burt, A. L. Rogach, A. Kornowski, A. EychmMller, H. Weller, *Pure Appl. Chem.* **2000**, *72*, 295.
4. P. M. Monk, R. J. Mortimer, D. R. Rosseinsky, *Electrochromism and Electrochromic Devices*, Cambridge University Press: New York, 2007.
5. (a) S. K. Deb, *Philos. Mag.* **1973**, *27*, 801. (b) S.-H. Lee, H. M. Cheong, J.-G. Zhang, A. Mascarenhas, D. K. Benson, S. K. Deb, *Appl. Phys. Lett.* **1999**, *74*, 242.
6. D. Davazoglou, A. Donnadieu, O. Bohnke, *Solar Energy Mater.* **1987**, *16*, 55.
7. (a) J. Fabian, H. Nakazumi, M. Matsuoka, *Chem. Rev.* **1992**, *92*, 1197. (b) M. Masaru, *Topics in Applied Chemistry. Infrared Absorbing Dyes*, Plenum Press: New York and London, 1990.
8. N. B. McKeown, *Phthalocyanine Materials: Synthesis, Structure and Function*, Cambridge University Press, Cambridge, 1998.
9. (a) K. D. Demadis, C. M. Hartshorn, T. J. Meyer, *Chem. Rev.* **2001**, *101*, 2655. (b) *Mixed Valence Compounds*, Brown, D. M., Ed.; D. Reidel: Dordrecht, Holland, 1980. (c) *Mixed Valence Systems: Applications in Chemistry, Physics*

and Biology, Prassides, K., Ed.; NATO ASI Series 343; Kluwer Academic Publishers: Dordrecht, The Netherlands, 1990. (d) R. J. Crutchley, *Adv. Inorg. Chem.* **1994**, *41*, 273.

10. *Ditholene Chemistry: Synthesis, Properties, and Applications*, Ed. Stiefel, E. I., John Wiley & Sons: Hoboken, New Jersey, 2004; in *Progress in Inorganic Chemistry*, Vol. 52, series edited by K. D. Karlin.

11. U. T. Mueller-Westerhoff, B. Vance, D. Ihl Yoon, *Tetrahedron*, **1991**, *47*, 909.

12. G. N. Schrauzer, V. P. Mayweg, *J. Am. Chem. Soc.* **1965**, *87*, 3585.

13. (a) S. I. Shupack, E. Billig, R. J. H. Clark, R. Williams, H. B. Gray, *J. Am. Chem. Soc.* **1964**, *86*, 4594. (b) M. J. Baker-Hawkes, E. Billig, H. B. Gray, *J. Am. Chem. Soc.* **1966**, *88*, 4870. (c) R. Williams, E. Billig, J. H. Waters, H. B. Gray, *J. Am. Chem. Soc.* **1966**, *88*, 43.

14. K. Ray, T. Weyhermüller, F. Neese, K. Wieghardt, *Inorg. Chem.* **2005**, *44*, 5345.

15. J. Campbell, D. A. Jackson, W. M. Stark, A. A. Watson, *Dyes Pigm.* **1991**, *15*, 15.

16. K. Y. Law, *Chem. Rev.* **1993**, *93*, 449.

17. Z. Peng, H. J. Geise, *Bull. Soc. Chim. Belg.* **1996**, *105*, 739.

18. I. H. Song, C. H. Rhee, S. H. Park, S. L. Lee, D. Grudinin, K. H. Song, J. Choe, *Org. Process Res. Dev.* **2008**, *12*, 1012.

19. S. Daehne, U. Resch-Genger, O. S. Wolfbeis, *Near-Infrared Dyes for High Technology Applications*, Kluwer Academic Publishers: Dordrecht, 1998.

20. J. Fabian, H. Nakazumi, M. Matsuoka, *Chem. Rev.* **1992**, *92*, 1197.

21. S.-i. Todoroki, S. Sakaguchi, K. Sugii, *Jpn. J. Appl. Phys., Part 1.* **1995**, *34*, 3128.

22. M. T. Harrison, S. V. Kershaw, M. G. Burt, A. L. Rogach, A. Kornowski, A. Eychmüller, H. Weller, *Pure Appl. Chem.* **2000**, *72*, 295.

23. S. Achilefu, R. B. Dorshow, J. E. Bugaj, R. Rajagopalan, *Invest. Radiol.* **2000**, *35*, 479.

24. D. J. Hawrysz, E. M. Sevick-Muraca, *Neoplasia.* **2000**, *2*, 388.

25. K. S. Schanze, J. R. Reynolds, J. M. Boncella, B. S. Harrison, T. J. Foley, M. Bouguettaya, T.-S. Kang, *Synth. Met.* **2003**, *137*, 1013.

26. F. Babler, US Patent 6989056 (2006).

27. S. Horiguchi, S. Ohira, Y. Abe, Y. Zama, H. Saikatu, EP 0211272 (1987).

28. (a) T. R. Sliwinski, R. A. Pipoly, R. P. Blonski, US Patent 20010022151 (2001). (b) L. Xie, Y. Ying, T. Ying, *J. Agric. Food. Chem.* **2007**, *55*, 4645.

chapter two

Molecular design and energy gap tuning

> Before God we are all equally wise—and equally foolish.

Albert Einstein (1879–1955)

The optical and electronic properties of organic molecules are determined by their energy gap, which is the energy separation between the highest occupied molecular orbital (HOMO) and lowest unoccupied molecular orbital (LUMO), also called the HOMO–LUMO gap (HLG). The molecules with extended conjugation or conjugated oligomers and polymers can be described by the band structures or the linear combination of molecular orbitals of multiple repeat units. The band gap between the valence and conduction band corresponds to the energy difference between the HOMO and LUMO and is relevant for optical absorption, photoluminescence, and other photophysical behaviors [1–3]. The band gap engineering is a key in the design of low energy gap organic materials and often involves chemical modification at the molecular level. Generally speaking, when tuning the energy gap of molecules, several factors should be considered, such as conjugation length, bond length alternation, and donor–acceptor charge transfer.

2.1 Effect of conjugation length

For a given π-conjugated unit or chromophore, extension of its conjugation length can certainly lead to a decrease in the HLG, although the scope and limitations of such a band gap reduction remains for further study and debate. Conceptually, the extension of a π-conjugation length involves transformation of a small π-conjugated unit into higher homologues by linear one-dimensional elongation and planar two-dimensional enlargement (Figure 2.1). The π-conjugated unit can be any small aromatic or heterocyclic molecules, such as benzene, naphthalene and thiophene, or even large molecules, such as porphyrins and phthalocyanines. By covalently linking or fusing π-conjugated units, one can extend the π-conjugation in a linear one-dimensional fashion. Another approach to the extension of π-conjugation is to enlarge the size of π-conjugated unit in plane in two

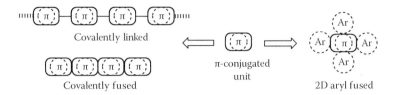

Figure 2.1 Transformation of a small π-conjugated unit into higher homologues by linear one-dimensional elongation and planar two-dimensional enlargement.

dimensions, usually by fusing additional π-conjugated aromatic (Ar) units that are ether identical to or different from the core unit in structure.

The classic example of one-dimensionally enchained π-conjugated systems is the stepwise buildup of higher homologues from benzene, going from biphenyl to terphenyl and to a homologous series of oligophenyls [4]. The longest wavelength of absorption maxima of ter-, quater-, quinque-, and sexi-phenyls are 279, 292, 299, and 308 nm, respectively [5]. The optical absorption energies, which are calculated from the longest wavelength of absorption maxima, were found to be in a linear relation with the inverse chain length ($1/n$, n: number of benzene rings) for the oligophenyls. However, the maximal absorption still falls within the visible region even as the number of phenyl rings increases, due to a result of the mutual distortion of the aromatic rings. Calculations show that biphenyl and terphenyl achieve their lowest energy conformations with torsion angles of 45° and 50°, respectively [6]. There is 23° mutual distortion of the phenylene subunits in oligophenyls [7]. Therefore, for the π-conjugated oligomers and polymers, the chain length at which the λ_{max} value no longer changes with added repeat units is defined as an "effective conjugation length" [8]. The effective conjugation length of oligophenyls is expected to be close to $n = 20$, based on the experimental results for its analogs [9].

Calculation shows that the energy levels of oligothiophenes vary with different numbers of the repeat units. With every extra thiophene unit being added, hybridization of the energy level continues, leading to a gradual decrease in the energy gap [10]. Although oligothiophenes absorb at a longer wavelength than oligophenyls, they still fall short of the NIR.

In comparison with covalently linking of the π-conjugated units via a single, double, or triple bond, covalently fusing of the π-conjugated units is a much more effective approach to lowering the band gap and shifting the absorption into the NIR region. For example, Zn(II)-porphyrin could be considered as a fairly large π-conjugated unit and absorbs around 420 nm. Osuka and Kim et al. successfully prepared the meso-meso-linked porphyrin arrays ($\mathbf{P_n}$) of up to 128 oligomers [11]. Although this rodlike molecule with a molecular length of about 108 nm is made up with the π-conjugated unit having a lower energy gap than benzene,

these arrays only absorb up to 600 nm, due to a nearly orthogonal conformation that tends to minimize the electronic interaction between the
neighboring porphyrins. To diminish or eliminate the electronic disrup-
tion of π-conjugation in the porphyrin arrays, the Osuka group developed
a facile synthetic protocol for fusing the porphyrin rings, which was also
utilized by other groups, to obtain fully conjugated porphyrin dimer, trimer, and even large tapes (**FP$_n$**) [12]. The photophysical properties of these
porphyrin tapes were studied in depth [13].

In sharp contrast, the fused porphyrins **FP$_n$** display a drastic redshift
in absorption spectra. The absorption spectra of the fused porphyrins
FP$_n$ in chloroform solutions exhibit three distinct bands (designated as
bands I, II, and III in Figure 2.2). Bands I are located in the visible region
(407–420 nm), which are basically the same positions as that of the
π-conjugated unit of Zn(II)-porphyrin; bands II cover the red (600 nm)
and NIR (900 nm) spectral regions, and bands III fall in the NIR and even
mid-IR regions. Bands I and II are assigned to allowed transitions along
the shorter and longer molecular axes of the rectangular fused porphyrin tapes, respectively. Thus, bands III correspond to a formally forbidden
transition such as Q-bands in the D4-symmetry Zn(II)-porphyrin mono-
mer. It is interesting to note that the absorption maxima of bands III (λ_{max})
correlate well with the number of porphyrins (N) in fused porphyrins or
the number of the repeat unit (n) in **FP$_n$** (N = n + 2, n = 0–10) (Figure 2.2).
The nearly straight line can be established and represented by the equation λ_{max} (nm) = 174.5N + 793 with a variation coefficient (r^2) of 0.995. Going
from dimer (N = 2) to dodecamer (N = 12), bands III shift progressively
into the infrared as follows: 1064 nm (dimer), 1333 nm (trimer), 1515 nm
(tetramer), 1667 nm (pentamer), 1852 nm (hexamer), 2222 nm (octamer),
and 2857 nm (dodecamer). Bands III of octamer and dodecamer of Zn(II)-
porphyrin tapes are located in the mid-IR region with their tails extending

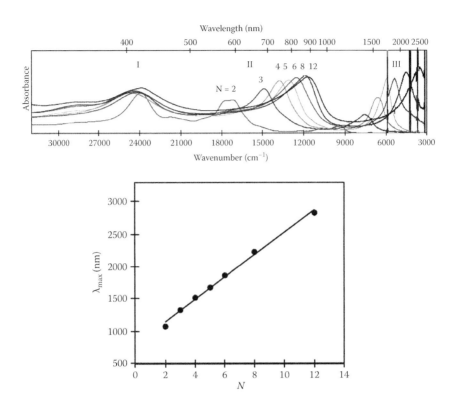

Figure 2.2 Absorption spectra of fused porphyrins **FP**$_n$ (labeled with the number of fused porphyrin rings) taken in chloroform at room temperature (top) and a plot of absorption maxima of bands III versus the numbers of porphyrins (*N*) in **FP**$_n$ (bottom). (Adapted with permission from A. Tsuda, A. Osuka, *Science,* **2001,** *293,* 79.)

beyond 3333 nm. The absorption data suggest a high possibility of further decrease in the excitation energy of bands III upon continuous elongation of the porphyrin arrays, since these known compounds have not shown the effect of the effective conjugation length. The case study on the fused porphyrin tapes clearly demonstrates the effectiveness of extension of π-conjugation by fusing the π-conjugated units in order to lower the band gap.

In an analogous fashion, by fusing benzene and naphthalene, polyacenes [14] and polyrylenes [15] can be obtained, respectively. In-depth theoretical and experimental studies on these fused aromatic compounds have made an enormous impact on the chemistry and applications of low band-gap polymers [16].

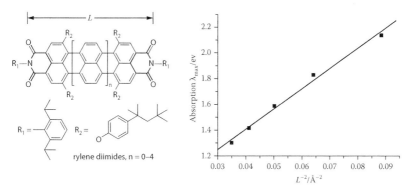

polyacenes   polyrylenes

For a series of rylene diimides, extension of π-conjugation or increase of the degree of annulation by going from perylene diimide to hexarylene diimide leads to a bathochromic shift of the absorption maximum (Figure 2.3). The maximum absorption energy is nearly linearly correlated to the inverse squared length (L) of π-aromatic system, where the length L is the distance between N-N in rylene diimides [17]. By having a different number of the fused naphthalene units in rylene diimides, the energy gap can be altered over a quite large range, from 1.25 to 2.15 eV. However, the energy gap of a conjugated π-system can only be lowered to a certain value by increasing the chain length, as the overall π-system is governed by the effective conjugated length [18–20].

Size enlargement by fusing π-conjugated moieties onto the core unit in two-dimensional fashion can lead to greater delocalization of the π-electrons and to a decrease in the HLG. The core unit can be as small as benzene and as large as porphyrin or other π systems. For example, the energy gap gradually decreases as the size of polycyclic aromatic hydrocarbons increases, starting from benzene to the one containing 222 carbons by the number of benzene units or sextet carbons (Figure 2.4). The term of *sextet* was initially used by Clar to describe the "benzene-like behavior"

Figure 2.3 Correlation between the maximum absorption energy and the inverse squared length of the π system. Plotted points represent the tetraphenoxy-substituted rylene dyes (left to right): hexarylene ($n = 4$), pentarylene ($n = 3$), quaterrylene ($n = 2$), terrylene ($n = 1$), and perylene ($n = 0$).

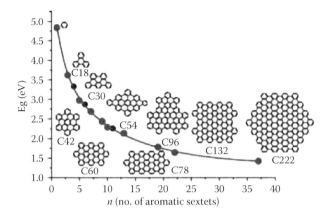

Figure 2.4 Correlation of the energy gap versus the number of aromatic sextets. (Adapted with permission from S. Müller, K. Müllen, *Phil. Trans. R. Soc. A.* **2007**, *365*, 1453.)

in polycycles that are only made from six-membered carbon rings [21]. In Clar's model, the π-electrons are assigned to specific six-membered rings in such a way that the largest possible number of π-electron sextets is formed. Accordingly, in polycyclic aromatic hydrocarbons shown in Figure 2.4, all π-electrons are distributable in closed sextets and the number of carbon atoms necessary is an integral multiple of six. Therefore, among some large polycyclic aromatic hydrocarbons, there exist so-called empty and full rings that refer to the central and outer rings, respectively. For example, in triphenylene there is one empty ring in the center and there are three full rings (or 3 aromatic sextets) around the central empty ring. In the case of the C222 molecule (Figure 2.4), there are 37 full aromatic rings or 37 sextets. The α-bands relate to the longest wavelength absorption of these large polycyclic aromatic hydrocarbons and can be used to calculate energy gaps.

By plotting the maximal peak value of first absorption α-bands of polycyclic aromatic hydrocarbons against the number (C_n) of carbon atoms, one can obtain a fairly linear line [22]. In dilute solution, the wavelength maximum of α-band increases linearly with C_n according to $\lambda_{max} = 280 + 2C_n$ and the spectral features become increasingly broadened. As the size of the polycyclic aromatic hydrocarbons increases, the first absorption band undergoes a pronounced bathochromic shift, from λ_{max} of 331 nm for C24 to 546 nm for C132 and to 724 nm for C222 (Figure 2.5). For a polycyclic benzenoid aromatic compound to show a maximal absorption at the telecommunication wavelength at 1600 nm, extrapolation indicates that such a NIR-absorbing compound is very large and needs to have 740 carbon atoms.

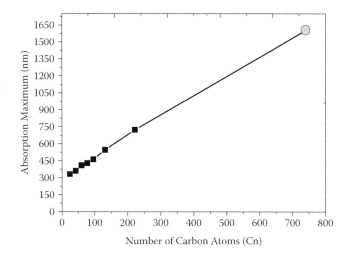

Number of Carbon Atoms (Cn)

Figure 2.5 Correlation of α-band absorptions of polycyclic benzenoid aromatic compounds versus the number of carbon atoms, extrapolated to include C740 with the estimated absorption maximum of 1600 nm.

These large, flat aromatic hydrocarbon π-systems represent a class of "molecular graphenes" and have received much interest for use as organic semiconductors [23]. The large π-expanded porphyrins, derived from the porphyrin core by fusing aromatic units, can be considered as a heterocyclic-containing molecular graphenes. A variety of mono-, di-, or triarene-fused porphyrins have been made and display long-wavelength bands in the NIR region, typically around 800–1000 nm [24–30]. By expanding the π system of porphyrin symmetrically in two dimensions, a class of highly symmetric, large tetraaryl-fused porphyrin molecules can be realized and are expected to have unusual optoelectronic properties, such as fused tetraanthracenylporphyrins that have been intrigued by people for more than 35 years [31]. In 2011, Anderson et al. successfully synthesized this compound, tetraanthracenyl-fused Ni-porphyrin (**Ni-FP**) [32]. Unlike many other unsymmetric aryl-fused porphrins, this porphyrin absorbs light intensively near the telecommunication band (λ_{max} = 1417 nm; ε = 1.2 × 10^5 M^{-1} cm^{-1}). In comparison with its meso-substituted tetraanthracenyl-porphyrin precursor (**Ni-P**), the longest absorption band for **Ni-FP** shifts to the red dramatically (Figure 2.6).

2.2   Effect of bond length alternation

In a typical π-conjugated system, the energy gap is originated from the bond length alternation between the double and single bond or so-called Peierls gap [33]. As Peierls predicts, the smaller bond length alternation

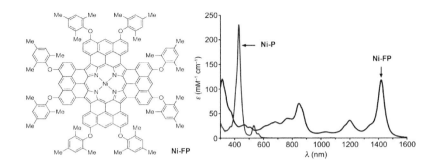

Figure 2.6 UV–vis–NIR spectra of meso-substituted tetraanthracenylporphyrin **Ni-P** and fused tetraanthracenylporphyrin **Ni-FP** in toluene. (Adapted with permission from N. K. S. Davis et al., *J. Am. Chem. Soc.* **2011**, *133*, 30. Copyright (2011) American Chemical Society.)

is, the lower energy gap of a conjugated compound will be. For example, polyacetylene would be a metallic conductor if the distance between all carbon atoms were identical [34,35]. Therefore, minimizing the bond length alternation is an important step toward the reduction of the energy gap of conjugated molecules.

The π-conjugated systems mainly include linear alternating polyene $(CH)_x$ and polyarylene compounds. The major difference between the former and the latter is the resonance energy of the monomer unit. In the former, the electron can delocalize along the chain if there is no deformation or defect; while in the latter system there exists a competition for π-electron between confinement within the rings and delocalization along the chain due to the aromaticity [36]. The aromaticity resonance energy, which is defined as the energy difference between the aromatic structure and a hypothetical reference consisting of isolated double bonds, is thus a measure for the stability and delocalization of a conjugated monomeric unit, which in turn dictates the final HLG of conjugated compounds and oligomers. As a result, the energy gap of aromatic molecules is larger than that of polyene molecules. Accordingly, the resonance energy of a conjugated building block should be taken into account in the molecular design.

Another difference between polyene and polyarylene π-conjugated systems is the nondegenerate ground state. The latter has the two limiting mesomeric forms, aromatic and quinoid, which are not energetically equivalent [37]. In most cases, the quinoid form has a smaller HLG [38]. Therefore, it is not so surprise for tetraphenylquinodimethane (**TPQDM**) [39] and diphenylindenofluorene (**DPIF**) [40] to be highly violet colored. The contribution of the quinoid state to the electronic ground state of the indenofluorene-containing polymer (**PolyIF**) is accounted for the observed low energy gap (*E*g ~ 1.55 eV) and the NIR absorption at 799 nm [41].

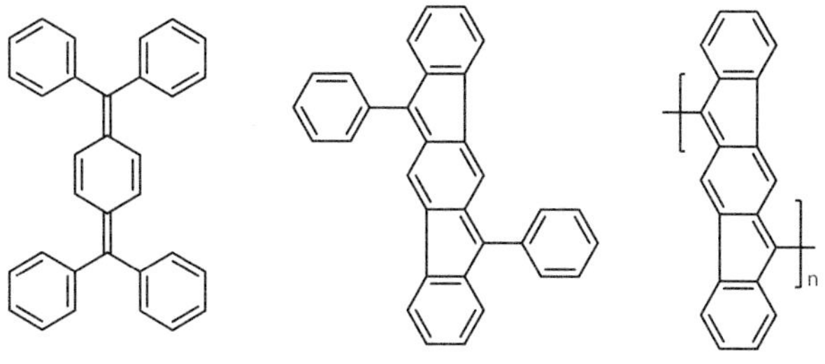

TPQDM **IF** **PolyIF**

Other quinoid-containing polymer examples include poly(*p*-phenylene), polypyrrole, and polythiophene. These conjugated long-chain macromolecules have a nondegenerate ground state since their ground state corresponds to a single geometric structure which, in their case, is aromatic-like. A quinoid-like resonance structure can be envisioned and has a lower ionization potential and a larger electron affinity than the aromatic structure by ab initio calculations (Figure 2.7) [42]. This explains why, on doping, the chain geometry in these compounds relaxes locally around the charges toward the quinoid structure that has a larger affinity for charges.

Therefore, generally speaking, an increase in the quinoid character in the designed molecules can be expected to effectively reduce the energy gap. At the same time, the quinoid structure represents the increased double-bond character of the bonds between the repeat units, leading to the reduction of bond length alternation. Accordingly, this approach has been routinely taken for designing and constructing NIR chromophores [43,44].

One exceptional case is worthwhile mentioning [12b]. As expected, the absorption of **FP2-CN** (λ_{max} = 958 nm; ε = 9.4 × 10^4 M^{-1}cm^{-1}) is sharper and more redshifted than that of nonplanar quinoidal porphyrin dimer **P2-CN** (λ_{max} = 780 nm; ε = 6.9 × 10^4 M^{-1}cm^{-1}). However, in comparison with the fused porphyrin dimer **FP2-Br** (λ_{max} = 1139 nm; ε = 1.1 × 10^4 M^{-1}cm^{-1}), the absorption of **FP2-CN** is blueshifted about 80 nm. The

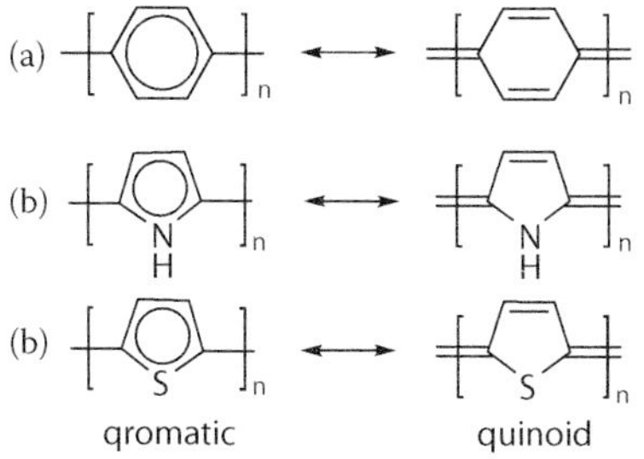

qromatic quinoid

Figure **2.7** Aromatic (ground-state) and quinoid-like geometric structures for (a) poly(*p*-phenylene), (b) polypyrrole, and (c) polythiophene.

unexpected blueshift on quinoidalization of **FP2-Br** to **FP2-CN** might be due to the greater bond-length alternation in quinoidalized porphyrins. Crystallographic studies revealed that the central $C_{meso} = C_{meso}$ bond length is 1.38 Å for **P2-CN** but only 1.43 Å for **FP2-CN**. Accordingly, the apparent double bond in **FP2-CN** is not significantly shorter than the formally single C_{meso}–C_{meso} bonds in Osuka's triply linked dimers [11].

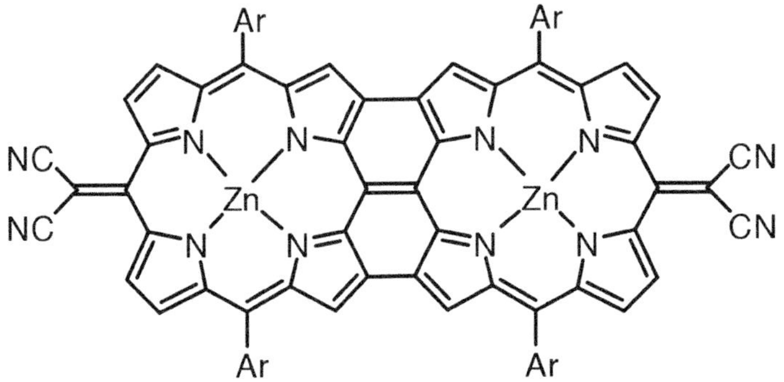

FP2-CN

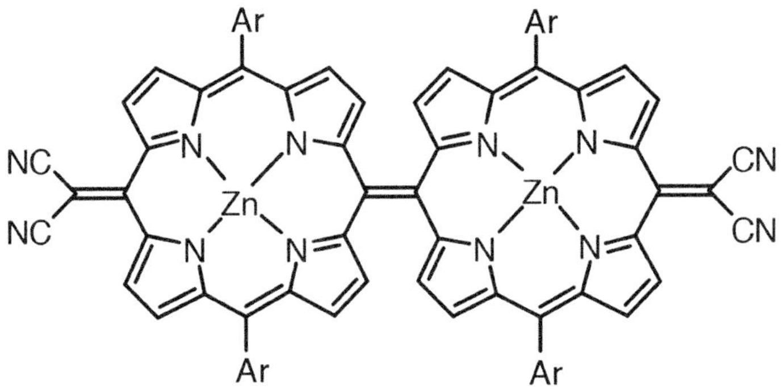

P2-CN

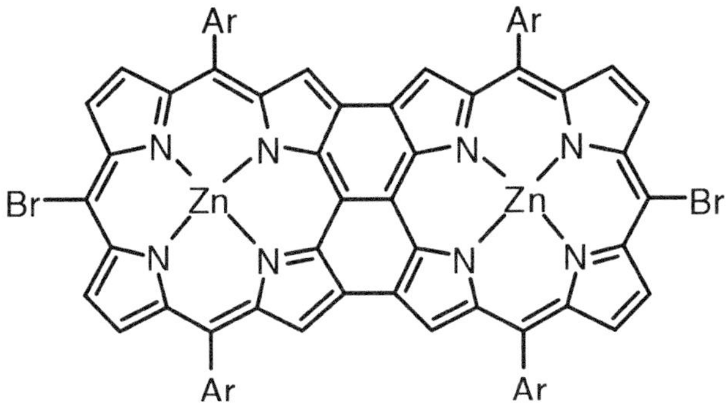

FP2-Br

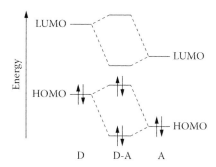

Figure 2.8 Interaction of energy levels of a donor (D) and acceptor (A) leading to a narrower HLG.

2.3   Effect of donor–acceptor charge transfer

Introduction of an electron donor (D) and an electron acceptor (A) in small molecules and polymers becomes a common strategy of lowering the energy gap [45]. On one hand, the D–A type of chromophores exhibit two resonance forms (D–A$^+$D = A$^-$), which gives rise to an increased double-bond character between the D and A units, thus reduce the bond length alternation, resulting in the decrease of the Peierls gap. On the other hand, hybridization of the energy levels of the donor and acceptor can raise the energetic level of the HOMO higher than that of donor and low the energetic level of LUMO than that of acceptor, leading to an unusually small HOMO–LUMO separation (Figure 2.8) [46,47].

For example, the unsubstituted bisazomethine compound **HH** exhibits a maximal absorption and emission at 440 and 534 nm, respectively. By substituting the nitro and diethylamino groups at the both ends, the push–push and pull–pull compounds (**DD** and **AA**) give a redshift of 110 nm and 41 nm in absorption, respectively, relative to compound **HH**. In the presence of one donor and one acceptor at each end, the push–pull compound **DA** has the most pronounced bathochromic shifts in absorption (148 nm) and emission (126 nm) [48].

HH: $R_1 = R_2 = H$	$\lambda_{max}^{abs} = 440$ nm, $\lambda_{max}^{pl} = 534$ nm
AA: $R_1 = R_2 = NO_2$	$\lambda_{max}^{abs} = 481$ nm, $\lambda_{max}^{pl} = 589$ nm
DD: $R_1 = R_2 = NEt_2$	$\lambda_{max}^{abs} = 555$ nm, $\lambda_{max}^{pl} = 659$ nm
DA: $R_1 = NEt_2; R_2 = NO_2$	$\lambda_{max}^{abs} = 558$ nm, $\lambda_{max}^{pl} = 660$ nm

In a D–A conjugated π-system, the donor and acceptor groups play a critical role in the band-gap tuning. A wide range of donor and acceptor molecules are known and have been used in the design and synthesis of low band-gap compounds and polymers. Considering the charge density as a measure for electron-withdrawing and accepting ability, the donor bearing a fully positive charge and the acceptor bearing a fully negative charge are considered to be the most powerful. Typical examples are ionic dyes. Crystal Violet is a cationic triphenylmethane dye with intense absorption in the visible region (λ_{max} = 588 nm), and some known triphenylmethane dyes absorb in a range of 705–830 nm [49].

Effective D–A charge transfer is essential to achieve the hybridization of the energy levels of the donor and acceptor, as evident by comparison of several trivinylogs of triphenylmethane dyes with the parent Crystal Violet and among themselves. They exhibit a very large bathochromic shift relative to Crystal Violet. Among them, the one bearing the ferrocenyl group as a donor (**CV-Fc**) absorbs at 1068 nm ($\varepsilon \sim 3.49 \times 10^4$ M^{-1} cm^{-1}) [49c], which closely matches those of NIR-absorbing cyanine and polymethine dyes (λ_{max} 850–1100 nm) [50]. Such a large bathochromic shift is, presumably, caused by lowering the energy of the metal-to-ligand charge-transfer (MLCT) transitions in the cationic ferrocenyl styryl chromophore of **CV-Fc**. Extended Hückel calculations have shown that, in this kind of D–A systems, an increase in the acceptor strength causes large bathochromic shifts due to considerable lowering of the already low-lying MLCT transitions [51]. The contribution of MLCT in inducing such a low-energy transition in **CV-Fc** is most evident when one compares its absorption data with the two structurally analogous cationic dyes containing methoxy (**CV-OR**) and dimethylamino (**CV-NR**) donors [52].

Given that all the cationic dyes contain a formal carbocation, the long-wavelength absorption at 1100 nm for a zwitterionic dye (**ZD**) having a rather simple structure is attributed to an unusual electron-withdrawing ability of the central mesoionic unit, aided by the two carbonyl groups of the lactone rings [53]. Such a zwitterionic unit acts as a "super acceptor group," since its ability as an acceptor in **ZD** clearly exceeds the acceptor properties of the most carbocations and neutral electron-withdrawing groups. Recent high-level calculations have shown conclusively that the NIR absorption in these molecules is not due to the increased donation and acceptance of electrons but is due to the diradicaloid nature of the central oxyallyl ring [54].

To design the low band-gap D–A chromophores and polymers, some powerful acceptor units have been realized either conceptually or practically and mainly include three series of strongly electron-deficient units: benzo-series, thieno-series, and bisbenzo-series. Depending on the atoms A or/and B used in the fused cyclic rings, the strength of these acceptors varies. When being linked with a variety of donors through a suitable π-conjugated spacer, these acceptors form a stockpile of building blocks for systematic construction of a wide range of low band-gap compounds and polymers.

A, B = O, NR, S, Se, Te

benzo-series   thieno-series   bisbenzo-series

In additional to the electron-donating groups such as NR_2, OR, and SR, a large electron-rich aromatic unit, can be used as a donor. The electron-donating character of thiophene- and fused thiophene-based derivatives has long been recognized and used to construct D–A compounds and polymers that usually absorb in the visible region. To red-shift the absorption, one needs to make thiophene-based donors more electron rich. By bridging the two thiophene rings or fusing the thiophene or aromatic unit to thiophene, one can obtain electron-rich thiophene-containing donors as shown in Figure 2.9 [55–57]. These donors are all considered to be fused thiophenes of three types: (1) thienothiophenes (TT) including thieno-[3,4-*b*]thiophene (TT1), thieno[3,2-*b*]thiophene (TT2), and thieno[2,3-*b*]thiophene (TT3); (2) benzodithiophenes (BDT) including benzo[1,2-*b*:4,5-*b*0]dithiophene (BDT1) and benzo[2,1-*b*:3,4-*b*0]dithiophene (BDT2); (3) β,β′-bridged bithiophenes (BBT) including dithienothiophenes

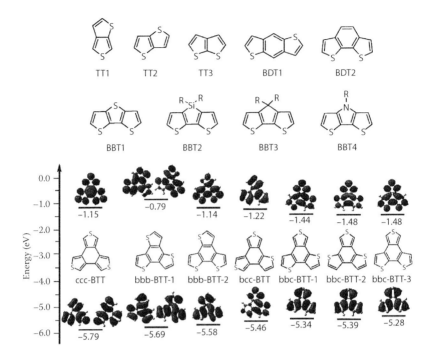

Figure 2.9 Molecular structures of three types of thiophene-fused donors (top) and seven benzotrithiophene isomers (bottom), along with the calculated frontier orbitals (DFT-B3LYP6-31G*). (Adapted with permission from X. Guo et al., *Org. Lett.* **2011**, *13*, 6062. Copyright (2011) American Chemical Society.)

(BBT1), dithieno[3,2-*b*:20,30-*d*]silole (BBT2), cyclopentadithiophene (BBT3), and dithieno[3,2-*b*:20,30-*d*]pyrrole (BBT4).

By cyclizing or fusing a terthiophene, one obtains a group of isomeric benzotrithiophenes (BTT) that all have a sulfur-rich, planar, and extended π-system (Figure 2.9) [58,59]. Among seven possible isomeric members of the BTT family, DFT calculations indicate that **bbc-BTT-3** possesses a higher HOMO level than all other isomers and is an electron-rich donor unit [59]. Further experimental work confirmed the theoretical predication, as evident by the measured absorption maximum close to 400 nm for the only alkyl-substituted **bbc-BTT-3**.

The spacer unit between the donor and acceptor also plays an important role in tuning the energy gap of D–π–A compounds, as it can alter the interaction of electronic orbitals of donor and acceptor [60]. The electronic and geometric nature of the spacer can affect the charge transfer and thus the overall energy gap of D–A chromophores. For example, the spacer effect in three structurally related chromophores, **DT-B, DT-N,** and **DT-A** in Figure 2.10, was studied in detail. These chromophores contain two

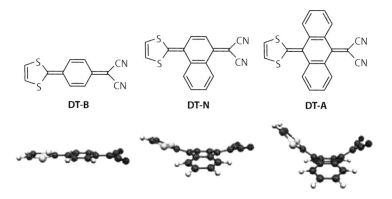

Figure 2.10 D–A chromophores **DT-B**, **DT-N**, and **DT-A** and their calculated geometries.

proaromatic moieties, namely, a 1,3-dithiole donor unit and a spacer incorporating a π-quinoid fragment [61].

By changing the spacer molecule from benzene to naphthalene to anthracene, there is a marked hypsochromic shift of the lowest energy absorption in dichloromethane [62], which correlates well to the increased HOMO–LUMO gaps calculated by DFT (2.42 eV for **DT-B**, 2.57 eV for **DT-N**, and 2.74 eV for **DT-A**). A tentative explanation to this behavior could rely on the different geometry displayed by these chromophores:

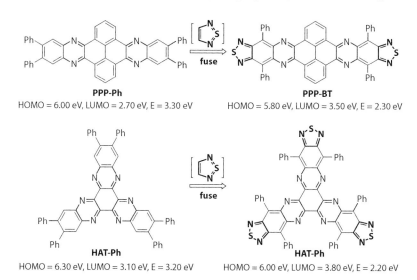

Figure 2.11 Computational energy levels of heterocyclic cores and thiadiazole-fused chromophores.

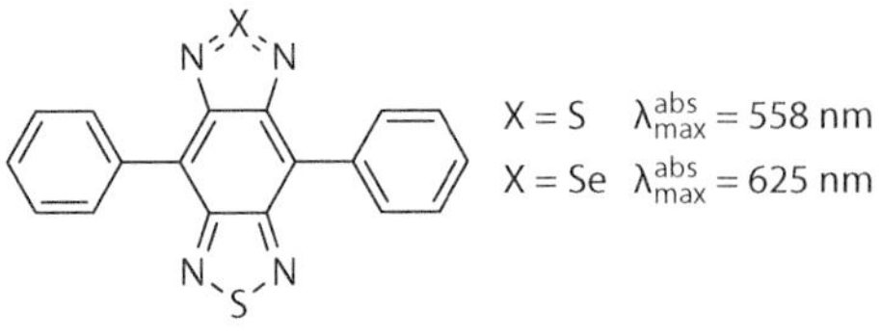

X = S λ_{max}^{abs} = 558 nm
X = Se λ_{max}^{abs} = 625 nm

Figure 2.12 Polar effect of heavy atoms, Se versus S.

while **DT-B** is completely planar (C_{2V} symmetry), there is some folding in **DT-N**. The shape of **DT-A** adopts a butterfly-like geometry, diminishing an effective charge transfer. As a consequence, the less proaromatic spacer in **DT-A** favors the neutral form over the zwitterionic one in the ground state. The different charge transfer of these compounds is also reflected in the calculated ground-state dipole moments that decrease in the order **DT-B** > **DT-N** > **DT-A**.

Finally, the D–A charge-transfer interaction can be further enhanced by attaching multiple electron-donating groups at the appropriate positions of the acceptor or with different donor–acceptor topology [63]. For example, **PPP-Ph** and **HAT-Ph** (Figure 2.11) were reported to have the energy gaps of 2.78 and 2.85 eV, or HOMO/LUMO energies of 5.92/3.25 and 6.37/3.70 eV, respectively [64,65]. The calculations indicate that **PPP-Ph** and **HAT-Ph** have a similar gap, 3.30 and 3.20 eV, respectively. By fusing a thiadiazole unit onto PPP and HAT cores, the calculations show a large decrease in LUMO levels (3.50 and 3.80 eV from 2.70 and 3.10 eV, respectively), resulting in a significant decrease in the energy gap for hypothetic **PPP-BT** (2.30 eV) and **HAT-BT** (2.20 eV).

2.4   Other effects

The effects on the energy gap tuning as discussed earlier are only applied to the individual molecules and oligomers without consideration of any intermolecular interaction. The intermolecular interaction such as hydrogen bonding, molecular stacking, and charge transfer also cause the change in the band gap for the molecules in the solid state [66,67].

The polar effect caused by the heavy atoms on the energy gap should also be considered when designing NIR chromophores. The well-known cases of lowering the energy gap typically involve the replacement of sulfur and oxygen atoms with heavier ones such as selenium and tellurium in a conjugated system [68]. A redshift of 67 nm in absorption maximum was observed in a simple diphenylbenzobisthiadiazole compound when the sulfur is replaced with the selenium (Figure 2.12) [69].

References

1. H. A. M. van Mullekom, J. A. J. M. Vekemans, E. E. Havinga, E. W. Meijer, *Mater. Sci. Eng. R.* **2001**, *32*, 1.
2. Z. Li, H. Meng, *Organic Light-Emitting Materials and Devices*, CRC Press: New York, 2007.
3. C. Winder, N. S. Sariciftci, *J. Mater. Chem.* **2004**, *14*, 1077.
4. Z. Xu, J. S. Moore, *Acta Polym.* **1994**, *45*, 83.
5. J. Grimme, M. Kreyenschmidt, F. Uckert, K. Müllen, U. Scherf, *Adv. Mater.* **1995**, *7*, 292.
6. (a) H. Cailleaau, J. L. Baudour, J. Meinnel, A. Dworkin, F. Moussa, C. M. E. Zeyen, *Faraday Discuss. Chem. Soc.* **1980**, *69*, 7. (b) S. Tsuzuki, K. Tanabe, *J. Phys. Chem.* **1991**, *95*, 139.
7. R. L. Elsenbaumer, L. W. Shdcklette, in *Handbook of Conducting Polymer*, Ed. T. A. Skotheimj, Marcel Dekker: New York 1986, Vol. I, p. 213.
8. (a) R. E. Martin, F. Diederich, *Angew. Chem. Int. Ed. Engl.* **1999**, *38*, 1350. (b) H. Meier, U. Stalmach, H. Kolshorn, *Acta Polym.* **1997**, *48*, 379. (c) B. B. Xu, J. H. Zhang, Z. H. Peng, *Synth. Met.* **2000**, *113*, 35. (d) S. Setayesh, D. Marsitzky, K. Müllen, *Macromolecules*, **2000**, *33*, 2016. (e) M. S. Wong, Z. H. Li, M. F. Shek, K. H. Chow, Y. Tao, M. D'Iorio, *J. Mater. Chem.* **2000**, *10*, 1805. (f) M. Remmers, B. Muller, K. Martin, H. J. Rader, W. Kohler, *Macromolecules,* **1999**, *32*, 1073.
9. J. Grimme, M. Kreyenschmidt, F. Uckert, K. Müllen, U. Scherf, *Adv. Mater.* **1995**, *7*, 292.
10. U. Salzner, J. B. Lagowski, P. G. Pickup, R. A. Poirier, *Synth. Met.* **1998**, *96*, 177.
11. (a) N. Aratani, A. Osuka, Y. H. Kim, D. H. Jeong, D. Kim, *Angew. Chem. Int. Ed.* **2000**, 39, 1458. (b) A. Tsuda, H. Furuta, A. Osuka, *J. Am. Chem. Soc.* **2001**, *123*, 10304.
12. (a) A. Tsuda, A. Osuka, *Science,* **2001**, *293*, 79. (b) I. M. Blake, A. Krivokapic, M. Katterle, H. L. Anderson, *Chem. Commun.* **2002**, 1662.
13. (a) H. S. Cho, D. H. Jeong, S. Cho, D. Kim, Y. Matsuzaki, K. Tanaka, A. Tsuda, A. Osuka, *J. Am. Chem. Soc.* **2002**, *124*, 14642. (b) D. Kim, A. Osuka, *Acc. Chem. Res.* **2004**, *37*, 735.
14. (a) V. R. Sastri, R. Schulman, D. C. Robert, *Macromolecules,* **1982**, *15*, 939. (b) H. Hart, J. Luo, *J. Org. Chem.* **1987**, *30*, 4833.
15. (a) Z. G. Soos, D. Mukhopadhyay, A. Painelli, A. Girlando, *Handbook of Conducting Polymers*, 2nd ed.; Marcel Decker: New York, 1998. (b) Z. Igbal, D. M. Ivory, J. Marti, J. L. Brédas, R. H. Baughman, *Mol. Cryst. Liq. Cryst.* **1985**, *118*, 103. (c) M. L. Kaplan, P. H. Schmidt, C. H. Chen, J. M. W. Walsh, *Appl. Phys. Lett.* **1980**, *36*.
16. (a) J. L. Brédas, R. H. Baughman, *J. Chem. Phys.* **1985**, *83*, 1316. (b) M. Pommerantz, R. Cardona, P. Rooney, *Macromolecules,* **1989**, *22*, 304. (c) P. M. Lahti, J. Obrzut, F. E. Karasz, *Macromolecules,* **1987**, *20*, 2023.
17. N. G. Pschirer, C. Kohl, F. Nolde, J. Qu, K. Müllen, *Angew. Chem. Int. Ed.* **2006**, *45*, 1401.
18. J. L. Bredas, R. Silbey, D. S. Boudreaux, R. R. Chance, *J. Am. Chem. Soc.* **1983**, *105*, 6555.
19. J. Grimme, M. Kreyenschmidt, F. Uckert, K. Müllen, U. Scherf, *Adv. Mater.* **1995**, *7*, 292.

20. J. Guay, P. Kasai, A. Diaz, R. Wu, J. M. Tour, L. H. Dao, *Chem. Mater.* **1992**, *4*, 1097.
21. (a) E. Clar, *The Aromatic Sextet*, John Wiley and Sons: London, 1972. (b) E. Clar, *Ber. Dtsch. Chem. Ges.* **1936**, *69*, 607.
22. (a) M. G. Debije, J. Piris, M. P. de Haas, J. M. Warman, Z. Tomovic, C. D. Simpson, M. D. Watson, K. Müllen, *J. Am. Chem. Soc.* **2004**, *126*, 4641. (b) M. D. Watson, A. Fechtenkötter, K. Müllen, *Chem. Rev.* **2001**, *101*, 1267.
23. (a) S. Müller, K. Müllen, *Phil. Trans. R. Soc. A.* **2007**, *365*, 1453. (b) L. Zhi, K. Müllen, *J. Mater. Chem.* **2008**, *18*, 1472.
24. (a) T. D. Lash, T. M. Werner, M. L. Thompson, J. M. Manley, *J. Org. Chem.* **2001**, *66*, 3152. (b) H. S. Gill, M. Harmjanz, J. Santamaría, I. Finger, M. Scott, J. *Angew. Chem., Int. Ed.* **2004**, *43*, 485. (c) O. Yamane, K. Sugiura, H. Miyasaka, K. Nakamura, T. Fujimoto, K. Nakamura, T. Kaneda, Y. Sakata, M. Yamashita, *Chem. Lett.* **2004**, *33*, 40.
25. (a) N. Nakamura, N. Aratani, H. Shinokubo, A. Takagi, T. Kawai, T. Matsumoto, Z. S. Yoon, D. Y. Kim, T. K. Ahn, D. Kim, A. Muranaka, N. Kobayashi, A. Osuka, *J. Am. Chem. Soc.* **2006**, *128*, 4119. (b) K. Kurotobi, K. S. Kim, S. B. Noh, D. Kim, A. Osuka, *Angew. Chem. Int. Ed.* **2006**, *45*, 3944.
26. S. Hayashi, M. Tanaka, Hayashi, S. Eu, T. Umeyama, Y. Matano, Y. Araki, H. Imahori, *J. Phys. Chem. C* **2008**, *112*, 15576.
27. (a) N. K. S. Davis, M. Pawlicki, H. L. Anderson, *Org. Lett.* **2008**, *10*, 3945. (b) N. K. S. Davis, A. L. Thompson, H. L. Anderson, *Org. Lett.* **2010**, *12*, 2124.
28. C. Sooambar, V. Troiani, C. Bruno, M. Marcaccio, F. Paolucci, A. Listorti, A. Belbakra, N. Armaroli, A. Magistrato, R. De Zorzi, S. Geremia, D. Bonifazi, *Org. Biomol. Chem.* **2009**, *7*, 2402.
29. C. J. Jiao, K. W. Huang, Z. P. Guan, Q. H. Xu, J. S. Wu, *Org. Lett.* **2010**, *12*, 4046.
30. V. V. Diev, K. Hanson, J. D. Zimmerman, S. R. Forrest, M. E. Thompson, *Angew. Chem. Int. Ed.* **2010**, *49*, 5523.
31. T. F. Yen, *The Role of Trace Metals in Petroleum*, Ann Arbor Science Publishers: Ann Arbor, MI, 1975.
32. N. K. S. Davis, A. L. Thompson, H. L. Anderson, *J. Am. Chem. Soc.* **2011**, *133*, 30.
33. R. E. Peierls, *Quantum Theory of Solids*, Oxford University Press: London, 1956.
34. L. M. Tolbert, *Acc. Chem. Res.* **1992**, *25*, 561.
35. T. A. Albright, J. K. Burdett, M.-H. Whangbo, in *Orbital Interactions in Chemistry*, Wiley: New York, 1985, p. 229.
36. V. Hernandez, C. Castiglioni, M. Del Zoppo, G. Zerbi, *Phys. Rev. B.* **1994**, *50*, 9815.
37. Y.-S. Lee, M. Kertesz, *J. Chem. Phys.* **1988**, *88*, 2609.
38. J. L. Bredas, *J. Chem. Phys.* **1985**, *82*, 3808.
39. M. Schoenberg, *J. Chem. Soc.* **1935**, 1403.
40. U. Scherf, *Synth. Met.* **1993**, *55*, 767.
41. H. Reisch, U. Wiesler, U. Scherf, N. Tuytuylkov, *Macromolecules*, **1996**, *29*, 8204.
42. J. L. Brédas, B. Thémans, J. G. Fripiat, J. M. André, R. R. Chance, *Phys. Rev. B: Condens. Matter*, **1984**, *29*, 6761.
43. G. Qian, B. Dai, M. Luo, D. Yu, J. Zhan, Z. Zhang, D. Ma, Z. Y. Wang, *Chem. Mater.* **2008**, *20*, 6208.

44. M. Karikomi, C. Kitamura, S. Tanaka, Y. Yamashita, *J. Am. Chem. Soc.* **1995**, *117*, 6791.
45. E. E. Havinga, W. Hoeve, H. Wynberg, *Polym. Bull.* **1992**, *29*, 119.
46. G. Brocks, A. Tol, *J. Phys. Chem.* **1996**, *100*, 1838.
47. G. Brocks, A. Tol, *Synth. Met.* **1996**, *76*, 213.
48. S. Dufresne, M. Bourgeaux, W. G. Skene, *J. Mater. Chem.* **2007**, *17*, 1166.
49. (a) H. Schmidt, R. Wizinger, *Liebigs Ann. Chem.* **1959**, *623*, 204. (b) S. Akiyama, S. Nakatsuji, K. Nakashima, M. Watanabe, H. Nakazami, *J. Chem. Soc. Perkin Trans. 1*, **1988**, 3155. (c) S. Sengupta, S. K. Sadhukhan, *J. Mater. Chem.* **2000**, *10*, 1997.
50. (a) J. Fabian, H. Nakazumi, M. Matsuoka, *Chem. Rev.* **1992**, *92*, 1197. (b) W. B. Tuemmler and B. S. Wildi, *J. Am. Chem. Soc.* **1958**, *80*, 3772. (c) F. M. Hamer, in *Chemistry of Heterocyclic Compounds*, Wiley Interscience: New York, 1964, Vol. 18. (d) M. Zollinger, *Colour Chemistry. Synthesis, Properties and Applications of Organic Dyes and Pigments*, VCH: Weinheim, 1987.
51. (a) M. L. H. Green, S. R. Marder, M. E. Thompson, J. A. Bundy, D. Bloor, P. V. Kolinsky, R. J. Jones, *Nature*, **1987**, *330*, 360. (b) J. C. Calabrese, L.-T. Cheng, J. C. Green, S. R. Marder, W. Tam, *J. Am. Chem. Soc.* **1991**, *113*, 7227. (c) H. E. Bunting, M. L. H. Green, S. R. Marder, M. E. Thompson, D. Bloor, P. V. Kolinsky, P. V. Jones, *Polyhedron*, **1992**, *11*, 1489.
52. S. Sengupta, S. K. Sadhukhan, *J. Chem. Soc. Perkin Trans. 1*, **2000**, 4332.
53. M. Tian, S. Tatsuura, M. Furuki, I. Iwasa, L. S. Pu, *J. Am. Chem. Soc.* **2003**, *125*, 348.
54. (a) H. Langhals, *Angew. Chem. Int. Ed.* **2003**, *42*, 4286. (b) K. Yesudas, K. Bhanuprakash, *J. Phys. Chem. A*, **2007**, *111*, 1943. (c) A. L. Puyad, Ch. Prabhakar, K. Yesudas, K. Bhanuprakash, V. J. Rao, *J. Mol. Struc. THEOCHEM*, **2009**, *904*, 1.
55. (a) Y. Zou, A. Najari, P. Berrouard, S. Beaupré, B. R. Aïch, Y. Tao, M. Leclerc, *J. Am. Chem. Soc.* **2010**, *132*, 5330. (b) E. Zhou, M. Nakamura, T. Nishizawa, Y. Zhang, Q. Wei, K. Tajima, C. Yang, K. Hashimoto, *Macromolecules* **2008**, *41*, 8302. (c) H. Bronstein, Z. Chen, R. S. Ashraf, W. Zhang, J. Du, J. R. Durrant, P. S. Tuladhar, K. Song, S. E. Watkins, Y. Geerts, M. M. Wienk, R. A. J. Janssen, T. Anthopoulos, H. Sirringhaus, M. Heeney, I. McCulloch, *J. Am. Chem. Soc.* **2011**, *133*, 3272.
56. (a) X. Zhan, Z. Tan, B. Domercq, Z. An, X. Zhang, S. Barlow, Y. Li, D. Zhu, B. Kippelen, S. R. Marder, *J. Am. Chem. Soc.* **2007**, *129*, 7246. (b) M. Zhang, H. N. Tsao, W. Pisula, C. Yang, A. K. Mishra, K. Müllen, *J. Am. Chem. Soc.* **2007**, *129*, 3472.
57. (a) R. Rieger, D. Beckmann, W. Pisula, W. Steffen, M. Kastler, K. Müllen, *Adv. Mater.* **2010**, *22*, 83. (b) Y. Liang, D. Feng, Y. Wu, S.-T. Tsai, G. Li, C. Ray, L. Yu, *J. Am. Chem. Soc.* **2009**, *131*, 7792. (c) G. Lu, H. Usta, C. Risko, L. Wang, A. Facchetti, M. A. Ratner, T. J. Marks, *J. Am. Chem. Soc.* **2008**, *130*, 7670. (d) I. McCulloch, M. Heeney, C. Bailey, K. Genevicius, I. Macdonald, M. Shkunov, D. Sparrowe, S. Tierney, R. Wagner, W. Zhang, M. L. Chabinyc, R. J. Kline, M. D. McGehee, M. F. Toney, *Nature Mater.* **2006**, *5*, 328.
58. (a) H. Hart, M. Sasaoka, *J. Am. Chem. Soc.* **1978**, *100*, 4326. (b) A. Patra, Y. H. Wijsboom, L. J. W. Shimon, M. Bendikov, *Angew. Chem., Int. Ed.* **2007**, *46*, 8814. (c) R. Proetzsch, D. Bieniek, F. Korte, *Tetrahedron Lett.* **1972**,

543. (d) N. Jayasuriya, J. Kagan, *J. Org. Chem.* **1989**, *54*, 4203. (e) T. Kashiki, S. Shinamura, M. Kohara, E. Miyazaki, K. Takimiya, M. Ikeda, H. Kuwabara, *Org. Lett.* **2009**, *11*, 2473. (f) Y. Nicolas, P. Blanchard, E. Levillain, M. Allain, N. Mercier, J. Roncali, *Org. Lett.* **2004**, *6*, 273. (g) T. Taerum, O. Lukoyanova, R. G. Wylie, D. F. Perepichka, *Org. Lett.* **2009**, *11*, 3230. (h) T. Kashiki, M. Kohara, I. Osaka, E. Miyazaki, K. Takimiya, *J. Org. Chem.* **2011**, *76*, 4061. (i) C. B. Nielsen, J. M. Fraser, B. C. Schroeder, J. Du, A. J. P. White, W. Zhang, I. McCulloch, *Org. Lett.* **2011**, *13*, 2414.

59. X. Guo, S. Wang, V. Enkelmann, M. Baumgarten, K. Müllen, *Org. Lett.* **2011**, *13*, 6062.

60. S.-i. Kato, T. Matsumoto, T. Ishi-i, T. Thiemann, M. Shigeiwa, H. Gorohmaru, S. Maeda, Y. Yamashita, S. Mataka, *Chem. Commun.* **2004**, 2342.

61. R. Andreu, M. J. Blesa, L. Carrasquer, J. Garín, J. Orduna, B. Villacampa, R. Alcalá, J. Casado, M. C. R. Delgado, J. T. L. Navarrete, M. Allain, *J. Am. Chem. Soc.* **2005**, *127*, 8835.

62. R. Gompper, H.-U. Wagner, *Tetrahedron Lett.* **1968**, 165.

63. M. Luo, H. Shadnia, G. Qian, X. Du, D. Yu, D. Ma, J. S. Wright, Z. Y. Wang, *Chem. Eur. J.* **2009**, *15*, 8902.

64. J. Hu, D. Zhang, S. Jin, S. Z. D. Cheng, F. W. Harris, *Chem. Mater.* **2004**, *16*, 4912.

65. B. R. Kaafarani, L. A. Lucas, B. Wex, G. E. Jabbour, *Tetrahedron Lett.* **2007**, *48*, 5995.

66. P. M. Grant, I. P. Batra, *Solid State Commun.* **1979**, *29*, 225.

67. A. Mishra, R. K. Behera, P. K. Behera, B. K. Mishra, G. B. Behera, *Chem. Rev.* **2000**, *100*, 1973.

68. H. Ammar Aouchiche, S. Djennane, A. Boucekkine, *Synth. Met.* **2004**, *140*, 127.

69. Y. Yamashita, K. Ono, M. Tomura, S. Tanaka, *Tetrahedron*, **1997**, *53*, 10169.

chapter three

Near-infrared organic compounds

> All truths are easy to understand once they are discovered.

Galileo Galilei (1564–1642)

Organic compounds that absorb or emit in the near-infrared (NIR) spectral region typically contain a conjugated π-system as a chromophore, in which the π-electron should be effectively delocalized within the entire chromophoric entity. In this chapter, organic compounds containing the rylene, polymethine, and meropolymethine, donor–acceptor chromophores, and metal complex chromophores are discussed. In addition, a new class of selective NIR-absorbing colorless chromophores is presented.

3.1 NIR compounds containing rylene chromophores

The simplest and most representative of π-conjugated polyene system is *trans*-polyacetylene. However, due to its bond fixation, its absorption wavelength is converged to a certain value; about 600 nm even with a longer chain length [1]. Cyclic π-conjugated polyene is another example. The absorption maxima of cyclic polyene and cyclic polyene-yne can extend into the NIR region only when its conformation is rigid and locked [2], although its long wavelength absorbance is quite weak (log $\varepsilon \approx 3$). Structurally, benzene can be considered to be a cyclic polyene and then acenes are also chromophores that are structurally related to the polyene as they are derived by means of annulations of benzene. Owing to more effective π-electron delocalization, the low-energy absorption peaks are red shifted as the number of annulated benzene rings increases. However, higher homologous acenes show a limited bathochromic shift (e.g., pentacence, λ_{max}^{abs} = 700 nm) due to their insufficient stability. The annulated arenes derived formally by connecting all the peri-positions of naphthalene have been designated by Clar as rylenes or oligo(peri-naphthalene)s [3]. Rylenes are another form of the extended polyene system. Going from perylene to pentarylene, this series of soluble rylenes exhibits a gradual bathochromic shift toward the NIR region [4]. A penta(peri-naphthalene) or pentarylene derivative absorbs at 750 nm, implying that further extension of π-conjugation in higher rylenes should push the absorption deeper into the NIR region. Since most polyenes and polyene-containing compounds are practically difficult to obtain and unstable in air, they are, except for perylene and rylene diimides, not ideal candidates as NIR materials for applications at the wavelengths over 800 nm.

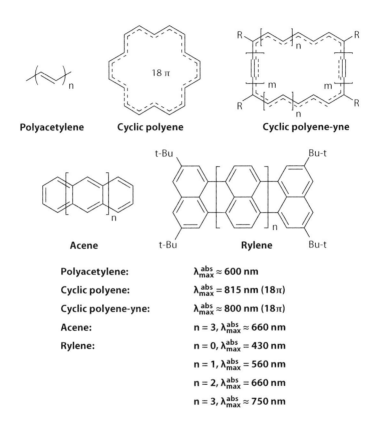

Polyacetylene:	$\lambda_{max}^{abs} \approx 600$ nm
Cyclic polyene:	$\lambda_{max}^{abs} = 815$ nm (18π)
Cyclic polyene-yne:	$\lambda_{max}^{abs} \approx 800$ nm (18π)
Acene:	$n = 3, \lambda_{max}^{abs} \approx 660$ nm
Rylene:	$n = 0, \lambda_{max}^{abs} = 430$ nm
	$n = 1, \lambda_{max}^{abs} = 560$ nm
	$n = 2, \lambda_{max}^{abs} = 660$ nm
	$n = 3, \lambda_{max}^{abs} \approx 750$ nm

Rylene diimides are a robust, versatile class of polycyclic aromatic electron-transporting materials with excellent thermal and oxidative stability, high electron affinity, and high electron mobility and therefore are promising candidates for a variety of organic electronics applications. Perylene diimides, as one of the smallest rylene derivatives, have a stable fluorescence property [5], and are readily available for use as color pigments [6], laser dyes [7], and photovoltaic materials [8]. Perylene diimide **1** is red, and thus this polyene chromophoric unit must be extended in order to realize the NIR absorption.

1, n = 0, R = H, $\lambda^{abs}_{max} \approx 590$ nm

2, n = 1, R = H, $\lambda^{abs}_{max} = 664$ nm

3, n = 2, R = H, $\lambda^{abs}_{max} = 762$ nm

4, n = 2, R = 4-t-butylphenoxy, $\lambda^{abs}_{max} = 781$ nm

5, n = 3, R = 4-t-octylphenoxy, $\lambda^{abs}_{max} = 877$ nm

6, n = 4, R = 4-t-octylphenoxy, $\lambda^{abs}_{max} = 953$ nm

In comparison with perylene diimide **1**, rylene diimides **2-65** have the extended π-conjugation along the long axis and absorb at 664–953 nm. Efficient and elegant syntheses of terrylene diimide **2** [9], quaterrylene diimide **3-4** [10], pentarylene diimide **5,** and hexarylene diimide **6** [11] by Müllen et al. warrant the subsequent study of optical and physical properties and exploration of potential applications. Quaterrylene diimide **3** was first synthesized in 1995 from perylenedicarboxylic imide in three steps, involving mono-bromination, dimerization by the Ni(0)-catalyzed Yamamoto coupling reaction, and final oxidative cyclization in the presence of base and an oxidizing agent (Scheme 3.1). Introduction of the four *t*-butylphenoxy groups at the bay-positions (β-position of imide groups) of the terminal naphthalene units impart good solubility to quaterrylene diimide **4**. Using the same coupling-oxidative cyclization strategy, soluble pentarylene diimide **5** and hexarylene diimide **6** were synthesized. With the extended π-conjugation over five and six fused naphthalene units, the maximal absorption extends over 800 nm. In the case of compound **6**, its absorption maximum reaches 953 nm, which represents a unique class of stable NIR-absorbing rylene dyes. The studies have shown that the maximal absorption energy of the rylene diimides is nearly linearly correlated to the inverse squared length of π system.

7, n = 0, R = H   8, n = 1, R = 4-*t*-butylphenoxy
$\lambda^{abs}_{max} = 778$ nm $\lambda^{abs}_{max} = 1106$ nm

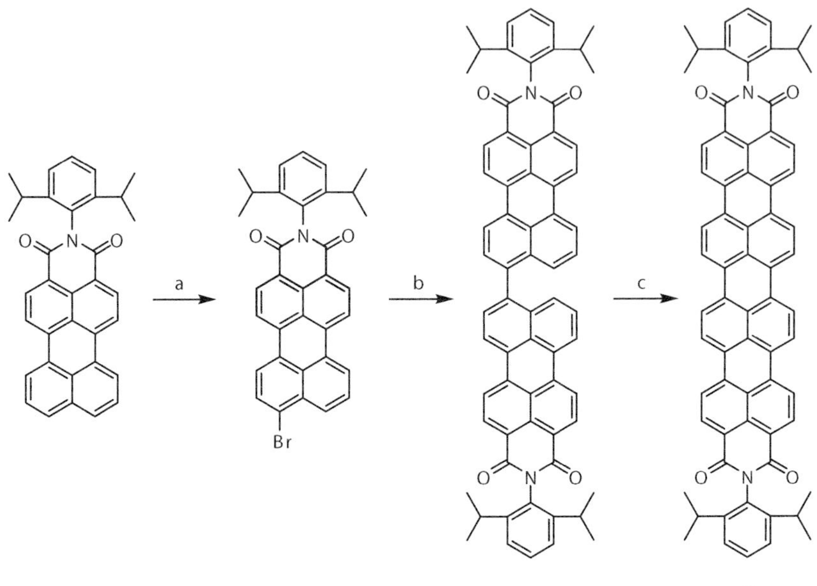

Scheme 3.1 Synthesis of quaterrylene diimide **3**. Reagents, conditions, and yields: (a) Br$_2$/C$_6$H$_5$Cl, 50°C. 98%; (b) Ni(cod)$_2$/DMF, 60°C, 2 d, 83%; (c) KOH/EtOH, 83%.

Another effective way to lower the energy gap is to insert a donor–acceptor (D–A) moiety into a polyene chromophoric unit. The structurally modified rylene dyes **7** and **8** have been successfully exploited by the Müllen group, which contain a D–A type of the diaminoanthraquinone unit within the extended rylene segment [12]. The synthesis is quite straightforward and involves the high-yield cross-coupling and oxidative cyclization reactions. Its absorption maximum at 778 nm can be attributed to the extension of the polyene length and, as well, the presence of a D–A unit. By analogy, compound **8** has even a lower energy gap with the absorption maximum at 1106 nm. These two examples demonstrate that insertion of the D–A unit into the extended polyene chromophore is an effective way to lower the energy gap.

9, R = CH(C_8H_17)_2
$$\lambda_{max}^{abs} = 692 \text{ nm}$$

10, R = CH(C_8H_17)_2
$$\lambda_{max}^{abs} = 778 \text{ nm}$$

$$\lambda_{max}^{pl} = 938 \text{ nm}$$

Attaching electron-donating groups to the bay position of perylene diimide is another useful method for shifting the absorption maxima to a longer wavelength. An increasing bathochromic shift is predicted if more donor groups are attached to the bay positions. It is found experimentally that introduction of the donor groups at other positions of perylene diimide has less effect on energy-gap tuning. Langhals and coworkers reported the perylene diimides containing the nitrogen donor groups (**9** and **10**) that have the absorption at the wavelengths (692 and 778 nm) longer than that of perylene diimide **1** and emit above 800 nm [13]. The neighboring donors joined to form six-membered rings can amplify the electron-donating effect. A variety of groups have been annulated onto the core bay of perylene diimides, such as aromatic rings [14] and heterocyclics [15], resulting in dramatic changes in the optical and electronic properties and, as well, intermolecular interactions [16].

11
$$\lambda_{max}^{abs} = 673 \text{ nm}$$
$$\lambda_{max}^{pl} = 694 \text{ nm}, \Phi_f = 0.26$$

12
$$\lambda_{max}^{abs} = 846 \text{ nm}$$

13
$$\lambda_{max}^{abs} = 760 \text{ nm}$$
$$\lambda_{max}^{pl} = 775 \text{ nm}, \Phi_f = 0.55$$

N-annulation at the bay positions was also done to some higher homologous rylenes, such as bis-N-annulated quarterrylenes **11** [17] and tri-N-annulated hexarylene **12** [18]. The key step in the synthesis of N-annulated rylenes involves the inter- or intramolecular oxidative coupling of the N-annulated perylene units by using a mixture of 2,3-dichloro-5,6-dicyano-1,4-benzoquinone (DDQ) and scandium trifluoromethanesulfonate [Sc(OTf)$_3$]. The mixed oxidizing agent of DDQ/Sc(OTf)$_3$ is particularly useful for the extension of electron-rich π-conjugated systems along the long molecular axis, such as triply linked oligoporphyrins [19], rylenes, and some polycyclic aromatics [20]. However, the C-C coupling of perylene units containing electron-withdrawing groups should preferably be carried out in an alkaline medium, typically in KOH or K$_2$CO$_3$ and ethanolamine as for the synthesis of quaterrylene diimide **13** [21]. In comparison with quaterrylene (λ_{max}^{abs} = 660 nm), N-annulation at the core does not induce a meaningful bathochromic shift in absorption (**11**, λ_{max}^{abs} = 673 nm). Only extension along the long molecular axis causes a significant redshift, as evident by N-annulated hexarylene **12**. Further redshift can be realized by introduction of the electron-withdrawing imide moiety at the end of the long molecular axis, as shown by bis-N-annulated quaterrylene diimide **13**. The absorption spectrum of compound **13** displays the longest wavelength band at 760 nm with a large molar extinction coefficient. However, when being compared with quaterrylene diimide 3 (λ_{max}^{abs} = 762 nm), the N-annulated analog **13** basically absorbs at the same wavelength but at a relatively shorter wavelength when being compared with analog **4** (λ_{max}^{abs} = 781 nm). Therefore, it can be concluded that the nitrogen in the pyrrole ring is weaker than the amino or even the phenoxy group in terms of the electron-donating strength.

14, λ_{max}^{abs} = 592 nm

λ_{max}^{pl} = 597, 650 nm, Φ_f = 0.10

15, λ_{max}^{abs} = 674 nm

λ_{max}^{pl} = 689, 756 nm, Φ_f = 0.55

Depending on the electron-donating nature, N-annulation at the bay region of rylenes tends to result in a larger redshift in absorption than the corresponding C-, S-, and Se-annulations [22]. Annulation by the ethylene or benzene units at the bay region of terrylene diimide afforded core-extended terrylene diimides **14** and **15**. These two compounds exhibited broad absorption spectra with high extinction coefficients, ranging from 300 to 700 nm. In comparison with terrylene diimide **2** (λ max = 664 nm), core extension along the short molecular axis as shown by **14** and **15** is a less effective way to achieve a significant bathochromic shift. Similarly, in the perylene system, N-, S-, and Se-annulations at the bay positions led to a slightly blueshift in absorption relative to the parent perylene (λ_{max} = 438 nm) by 13, 22, and 26 nm, respectively, due to the extended aromatic core along the short molecular axis [23]. A hypsochromic shift was even observed in a perylene diimide system upon the enlargement of the π-system along the short molecular axis [15a].

QR-DI-2N, λ_{max} = 752.0 nm

QR-DI-3N, λ_{max} = 689.9 nm

QR-DA-2N, λ_{max} = 742.9 nm

HR-DP-3N, λ_{max} = 911.9 nm

QR-AI-2N, λ_{max} = 794.5 nm

QR-AI-1N, λ_{max} = 833.8 nm

The quantum-chemical calculations of linear optical properties were carried out for a series of N-annulated rylenes and their mono- and diimides by means of the TD-DFT//B3LYP/6-31G* approach on the basis of the optimized geometry [24]. The effect of solvent (chloroform) was also taken into account at the same level. **QR-DI-2N** is virtually the same as compound **13,** and its calculated absorption maximum in chloroform (752.0 nm) matches well the experimental value (760 nm). A hypsochromic shift was again observed for **QR-DI-3N** (689.9 nm) when introducing an extra N-annulation at the core. **QR-DA-2N** is an analog of quarterrylene **11** (λ_{max} = 673 nm) and shows a redshift in absorption wavelength to 742.9 nm due to the presence of a stronger electron-donating diphenylamino group. The extension of the conjugated framework along the long molecular axis induces a large bathochromic shift of 169 nm for **HR-DP-3N** (λ_{max} = 911.9 nm) relative to that of **QR-DA-2N**. It is noteworthy that the λ_{max} value is bathochromically shifted from 742.9 nm for **QR-DA-2N** to 794.5 nm for **QR-AI-2N** when replacing one of the terminal groups with the imide unit. Interestingly, the calculation shows that the electronic densities of HOMO and LUMO are located at the center for the symmetrical molecules such as **QR-DA-2N** and **QR-DI-2N**. Such an electronic density distribution suggests that the $\pi-\pi^*$ transitions occur. However, for the unsymmetrical **QR-AI-2N** and **QR-AI-1N**, the charge transfer clearly takes place from the terminal groups to the center for the HOMO→LUMO transition. Accordingly, **QR-AI-2N** displays the lower energy gap than **QR-DA-2N** or **QR-DI-2N**. Removal of one N-annulated moiety from **QR-AI-2N** further lowers the energy gap of **QR-AI-1N** (λ_{max} = 833.8 nm), as a result of more effective charge transfer between the N-donor and the imido acceptor and reverse core extension along the short molecular axis.

3.2   NIR compounds containing polymethine and meropolymethine chromophores

Polymethines have been known as dyes since the early 1920s and found useful applications in photographic sensitization at wavelengths up to 1300 nm. This class of dyes continues receiving much interest owing to their intrinsic merits and diverse applications in bio-imaging [25], photovoltaics [26], light-emitting diodes [27], and nonlinear optics [28]. To date, a vast number of visibly colored polymethine and meropolymethine dyes have been documented and the structure–property relationships are reviewed in depth [29,30].

X = X′ (same hetero atom) : Polymethine dyes
X = NR, X′ = N⁺R₂ : Cyanine dyes
X = O⁻, X′ = O : Oxonole dyes

X ≠ X′ (different hetero atom) : Meropolymethine dyes
X = N, X′ = O : Merocyanine dyes

Merocyanine

Cyanine

Streptocyanines or open chain cyanines   Hemicyanines   Closed chain cyanines

Donor–acceptor-substituted polyenes containing same heteroatom (either nitrogen or oxygen) at their terminals are categorized into polymethine dyes; for those containing both nitrogen and oxygen as a donor–acceptor pair, they are categorized into meropolymethine dyes [29]. Merocyanine dyes have both nonpolar and zwitterionic resonance forms, and the spectral properties of merocyanines are very sensitive to the solvent polarity. In comparison, polymethine dyes are charged molecules, including cyanine, hemicyanine, and oxonol, and their spectral properties are not significantly influenced by the solvent polarity [31]. Merocyanine or hemicyanine dyes have been widely studied for their nonlinear optical (NLO) properties owing to their wide transparent range, high NLO coefficient, large molecular hyperpolarizability (β), and short response time [30]. Cyanine and merocyanine dyes have been extensively used for fluorescence labeling of protein, nucleic acid, and many others in various biological applications, because of their superb fluorescence brightness, high quantum yields and extinction coefficients, photostability, and long emission wavelengths without overlapping the cellular autofluorescence [32].

Naph Cy 7   **IR 1050**
λ_{max} (abs) = 750 nm λ_{max} (pl) = 830 nm (MeOH) λ_{max} (abs) = 1048 nm (EtOH)

The energy gap or optical properties of cyanine dyes can be tuned by structural modification and vary over a broad range. Structural modifications are usually aimed at either shifting the absorption to a longer wavelength or increasing the emission efficiency. For cyanine dyes, a redshift can be readily achieved by extending the π-conjugated chain length and, within a limit, every vinylene unit added leads to nearly a 100-nm bathochromic shift [33]. In addition, the molar absorption coefficients and oscillator strength of cyanine dyes increase noticeably as the chain length extends. The annulations of a benzene ring at the C-4 and C-5 positions of indolenine nuclei is another option to tune the emission maximum of cyanine dyes, which roughly results in 30~40 nm of bathochromic shift. A combination of both approaches can lead to a redshift into the NIR region (e.g., **Naph Cy 7**, λ_{max} = 750 nm). The tunability on emission wavelength of cyanine derivatives is based on the understanding of structure–photophysical property relationships, which allows the development of NIR fluorophores [30,34]. However, the modification through the chain extension is generally accompanied by a significant decrease in photo- and thermal stability, thus hampering their practical applications. Enhancement of the rigidity in the polymethine chain through alicyclic (e.g., cyclohexenyl) rings (e.g., **IR1050**, λ_{max} = 1048 nm) is very effective in improving their photostability [35] and applicability as NIR sensitizing dyes [29].

Polymethine dyes are fluorescent and their emission properties depend on the chain length and terminal groups. It was found by Ischenko et al. that the fluorescence efficiency reaches a maximum and then decreases going from a short to a long chain length [36]. Although some characteristics of most polymethine dyes, such as wavelength tunability, are favorable for use as NIR materials, their fluorescence quantum yields are still rather low, usually less than 15%, due to the chain flexibility that allows for photoisomerization in the excited state [37].

Borondipyrromethene

Aza-borondipyrromethene

Borondipyrromethenes (BODIPY) can be regarded as rigid, cross-conjugated cyanines. Introducing nitrogen at the C8-position in BODIPY results in analogous dyes, namely, aza-borondipyrromethenes (aza-BODIPY). By complexation with boron to lock up the chromophore conformation, the fluorescence quantum yield increases dramatically. The parent BODIPY has been structurally modified by many synthetic methods to yield a large number of BODIPY dyes that show the long-wavelength absorption and high-fluorescence quantum yield. The design strategy mainly aims at extending the conjugation length and lowering the resonance energy [38,39], introducing the rigid and fused rings [40–42] and attaching strong electron-donating groups to the various positions on BODIPY [43]. Several reviews have covered the subject of the structures and properties of BDPIPY dyes [44–47]. Thus, our attention is paid particularly to those that absorb and emit near and above the wavelength of 750 nm.

16, $\Phi_f = 0.29$ (CH$_2$Cl$_2$)

$\lambda_{max}^{abs} = 718$ nm

$\lambda_{max}^{pl} = 784$ nm

17, $\Phi_f = 0.20$ (CH$_2$Cl$_2$)

$\lambda_{max}^{abs} = 727$ nm

$\lambda_{max}^{pl} = 780$ nm

18, (CH$_2$Cl$_2$)

$\lambda_{max}^{abs} = 765$nm

$\lambda_{max}^{pl} = 827$ nm

Yu and coworkers reported a series of NIR fluorescent dyes (e.g., **16**) by elongating either one or two π-systems. The absorption and emission maxima are located at 718 and 784 nm, respectively. The solvatochromic effect was observed, indicating the existence of intramolecular charge transfer and explaining a low-fluorescence quantum yield ($\Phi_f = 0.1$) in polar solvents [48].

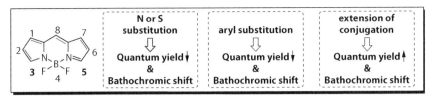

The extensive studies on the structure–property relationships of BODIPY reveal a number of important findings that are useful for design of NIR-absorbing and fluorescent BODIPY dyes. The C-3 and C-5 positions of BODIPY core skeleton is one of the most important points for the substitution to tune photophysical properties, including absorption and/or emission maxima as well as fluorescence quantum yields. First of all, the introduction of amine or sulfur-containing electron-donating group can lead to a significant bathochromic shift of both absorption and emission maxima along with a decrease in quantum yields relative to the parent BODIPY. The introduction of aryl substituents at the C-3 and C-5 positions also cause bathochromic shifts with reduced quantum yields. Only the vinyl substitution at the same positions can bring a bathochromic shift without a significant decrease in quantum yield. Therefore, the extension of conjugation length by the introduction of vinyl group at the C-3/C-5 positions is more effective for the bathochromic shift of the BODIPY system.

Compared to the introduction of various vinyl or aryl substituents at the BODIPY core skeleton, the extension of conjugation length can be more effectively achieved by fusion with aryl or heterocyclic groups. In fact, many research groups have applied the so-called aryl-ring fusion approach to the extension of the conjugation length while maintaining the structural rigidity in order to induce a bathochromic shift of absorption and emission maxima in given fluorophores [49–55]. Extension of the π-conjugation by fusion at the 1,2- and 6,7-positions of the two pyrrole moieties can lead to a noticeable bathochromic shift in absorption and emission maxima of BODIPY, as shown by compounds **17** [56] and **18** [57], whose absorption and emission maxima fall in the NIR region.

Fusion at the C-2/C-3 and C-5/C-6 positions with aryl or furan rings significantly affects the photophysical property and causes bathochromic shifts in both absorption and emission maxima compared with alkyl-substituted counter parts. These types of structural modifications often lead to dramatically enhanced fluorescence quantum yields and molar absorbance [58,59]. Most of the furan-fused BODIPY dyes exhibit a high molar absorbance ($\varepsilon = 2.53{\sim}3.16 \times 10^5$ $M^{-1}{\cdot}cm^{-1}$) and high quantum yields ($\Phi_F = 0.81{\sim}0.97$) and long emission wavelength (>650 nm).

Finally, fusion at C-1, C-7, and C-8 positions or the zigzag edge of BODIPY is another approach to the development of NIR fluorescent dyes. The Wu group successfully utilized the intramolecular oxidative cyclo-dehydrogenation reaction to fuse an electron-rich aromatic group onto BODIPY [60]. By this approach, anthracene-, perylene-, and porphyrin-fused BODIPY dyes with long-wavelength absorption/emission and high photostability have been successfully prepared [61]. Anthracene-fused BODIPY **20** was synthesized by intramolecular oxidative cyclodehydroge-nation of the C-8 substituted precursor **19** with excessive ferric chloride in 45% yield. The low yield was due to the subsequent intermolecular oxida-tive coupling to form the corresponding dimer **21** in 37% yield. In compar-ison with BODIPY **19**, the anthracene-fused BODIPY **20** and **21** exhibit a significant bathochromic shift and show a unique broad absorption band from 500 nm to 930 nm (Figure 3.1). Replacement of anthracene with por-phyrin, the porphyrin-fused BODIPY has a similar spectral feature as **20** and **21**. However, by fusion of two BODIPY molecules with one porphyrin, the corresponding double-fused BODIPY–porphyrin–BODIPY compound shows absorption deep beyond 1000 nm with an absorption maximum at 1040 nm ($\varepsilon = 68\,000\,M^{-1}\,cm^{-1}$), which is the longest absorption maximum ever observed for BODIPY derivatives to date [60].

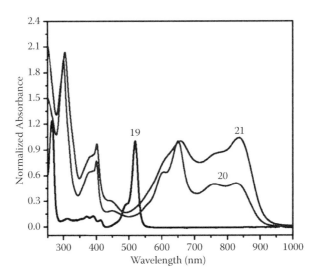

Figure 3.1 Absorption spectra of **19, 20,** and **21** in dichloromethane (1.1×10^{-5} M). The spectra were normalized at 650 nm for **20** and at 660 nm for **21**. (Adapted with permission from L. Zeng et al., *Org. Lett.* **2011**, *13*, 6026. Copyright (2011) American Chemical Society.)

22, $\Phi_f = 0.28$ (CHCl$_3$)
$\lambda^{abs}_{max} = 740$ nm
$\lambda^{pl}_{max} = 751$ nm

23, $\Phi_f = 0.29$ (CHCl$_3$)
$\lambda^{abs}_{max} = 736$ nm
$\lambda^{pl}_{max} = 748$ nm

24, (CHCl$_3$)
$\lambda^{abs}_{max} = 799$ nm
$\lambda^{pl}_{max} = 823$ nm

25, $\Phi_f = 0.05$ (CHCl$_3$)
$\lambda^{abs}_{max} = 774$ nm
$\lambda^{pl}_{max} = 815$ nm

Aza-BODIPY dyes were first reported in the 1940s and remained idle for a long time until 2000s [62]. Since 2002, largely led by the O'Shea group, the research on this class of dyes has been resurged [63]. Due to the presence of nitrogen lone pair of electrons in aza-BODIPY, the orbital levels of the cyanine framework are affected, resulting in a reduction of the HLG. They are typically red and NIR fluorophores with remarkable photophysical properties such as large extinction coefficients ($7.5–8.5 \times 10^5$ M^{-1} cm^{-1}) and high-fluorescent quantum yields above 700 nm [64]. The absorption of the aza-BODIPY-BF$_2$ chelates strongly depend on the sub-stituents on the aryl rings. Electron-donating groups on the para position of the aryl rings α to the pyrrole nitrogen tend to cause a large redshift, whereas electron-donating groups on the para position of the aryl rings β to the pyrrole nitrogen typically result in a smaller bathochromic shift. A series of new NIR aza-BODIPY dyes have been obtained by further energy gap engineering through introduction of the electron-donating groups [65] and stiffening the chromophore moieties [66]. For example, structurally rigid aza-BODIPY **22** was synthesized in two steps in 76.5% overall yield. It absorbs at 740 nm ($\varepsilon = 159000$ M^{-1}cm^{-1}) and emits at 751 nm ($\Phi_f = 0.28$) [67]. Substitution at the β position of the pyrrole does not affect much the optical properties of the aza-BODIPY dyes, as compound **23** having additional electron-rich anisole group absorbs at 736 nm with only a small hypsochromic shift relative to analogous **22** and has a slightly higher fluorescence quantum yield [67]. Compound **24**, synthesized in three steps, exhibits the absorption maximum at 799 nm with an extinction coefficient of 87000 M^{-1}cm^{-1} [67]. It is interesting to note that even though this structure contains a strong electron donor, its maximal absorption band shows little solvent dependence. The properties of unsymmetric dyes such as **25** are also highly dependent on the substituents. Introducing the electron rich para-piperidino group results in pushing the absorption to a quite long wavelength (774 nm for **25**), although it comes with a tradeoff with the decreased fluorescence quantum yield [67].

26a-f: R = F
27a-f: R = Ph

Pyrrolopyrrole cyanine (PPCy) is a relatively new class of dye that contains the pyrrolopyrrole core as a cyanine-type chromophore. PPCy are almost exclusively synthesized by condensation reaction of diketopyrrolopyrroles (DPPs) with α-substituted acetonitriles. Because of the low reactivity of the carbonyl group in DPPs, activation by $POCl_3$ is needed before reacting with the nitriles in a one-pot synthesis. The synthesis and optical properties of a series of PPCy dyes with various heterocyclic groups have recently been reported by Fischer and coworkers [68–70]. Due to the flexible and twisted structure, these PPCy compounds as the precursors to **26** and **27** do not fluoresce at all. Substitution of the protons in both N-H•••N bridges by BF_2 or BPh_2 locks all the chromophoric units in place and thus eliminate the torsion-induced radiationless S_1 decay [71,72], affording highly fluorescent dyes **26** and **27** [70]. All of the six derivatives of dyes **26** have the absorption maxima above 684 nm but below 800 nm (Table 3.1). Only compound **26f** containing the quinoxaline terminal groups in this series emits about 800 nm, presumably due to the lower LUMO level caused by the presence of the electron-withdrawing quinoxaline moiety. Compared to the BF_2 analogues **26**, the BPh_2–PPCy complexes **27** exhibit a bathochromic shift in optical properties, which is independent of the nature of the terminal heterocycle. This phenomenon is believed to be due to the different σ-inductive effect of BF_2 and BPh_2 groups on the chromophoric moiety. Due to this effect, the absorption and emission spectra of **27e** are a redshift about 65 nm with reference to **26e** and display the peak maxima at 819 nm and 831 nm, respectively. As expected, among all the quinoxaline derivatives, **27f** absorbs and emits at the longest wavelength. The maximum absorption and emission wavelengths of

Table 3.1 Optical properties of PPCy–BR_2 **26** and **27**

PPCy-BR_2	λ_{max}^{abs} [nm]	ε [M^{-1} cm^{-1}]	λ_{max}^{PL} [nm]	Φ_f
26a	684	125000	708	0.63
26b	690	135000	712	0.57
26c	707	145000	730	0.47
26d	732	190000	749	0.69
26e	754	205000	773	0.59
26f	789	210000	805	0.32
27a	737	163000	749	0.62
27b	747	164000	762	0.54
27c	763	169000	776	0.48
27d	790	221000	804	0.42
27e	819	256000	831	0.53
27f	864	261000	881	0.32

Note: In chloroform at room temperature.

27f reach to 864 nm and 881 nm, respectively. More importantly, at such a long wavelength, its fluorescence quantum yield is as high as 32%. In addition, these dyes show good thermal stability and photostability.

Squaraines belong to a class of polymethine dyes with resonance-stabilized zwitterionic structures and can be considered as a special kind of cyanine dyes. Similar to the latter, squaraines show an odd number of carbon atoms between the two nitrogen atoms fencing the conjugation along the polymethine chain. In general, squaraine dyes are close in optical properties to the cationic cyanine dyes. They show intense absorption and emission properties in the visible and NIR regions and have been widely used in imaging, nonlinear optics, photovoltaics, biological labeling, and photodynamic therapy. Introducing a strong electron donor or extending the π-conjugation length can lower the energy gap and shift the absorption of squaraines toward the NIR region. Using this approach, a large number of squaraines with NIR absorption and emission have been synthesized [73]. The relationship between the physical and optical properties and the chemical structures of squaraines and squaraine-containing polymers has been reviewed in detail [74].

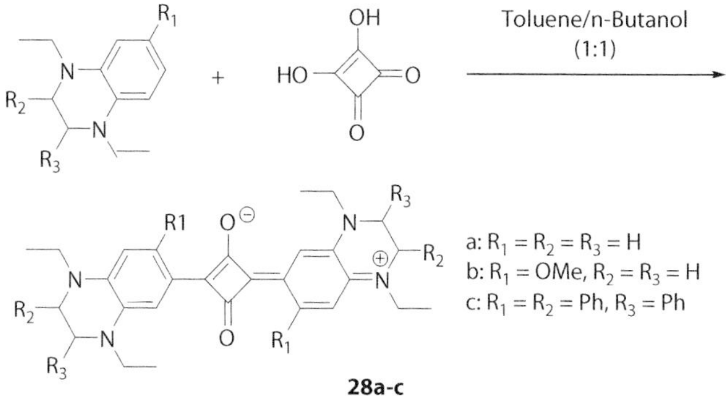

a: $R_1 = R_2 = R_3 = H$
b: $R_1 = OMe, R_2 = R_3 = H$
c: $R_1 = R_2 = Ph, R_3 = Ph$

28a-c

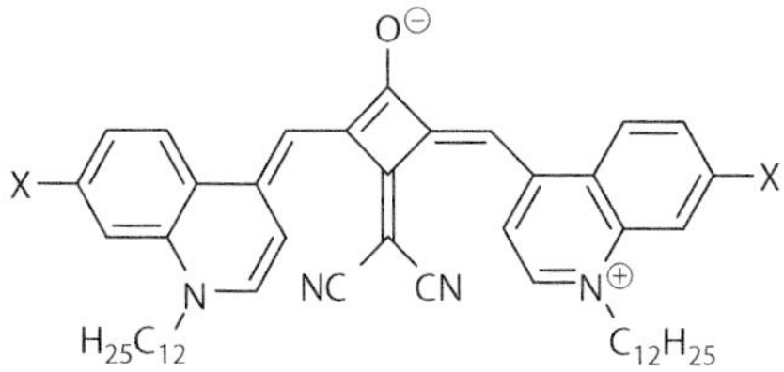

29a-d
a: X = H, b: X = Cl, c: X = Br, d: X = I

It is still worthwhile mentioning a report by Chandrasekaran et al. on a new series of NIR squaraine dyes **28a-c** having tetrahydroquinoxaline as an electron donor [75]. Three derivatives, which were readily synthesized by refluxing the tetrahydroquinoxaline derivatives with squaric acid in

Table 3.2 Optical properties of squaraines
28a-c and 29a-d[a]

	λ_{max}^{abs} [nm]	λ_{max}^{pl} [nm]	Φ_f
28a	711	790	9.6×10^{-3}
28b	732	823	3.0×10^{-3}
28c	717	774	2.9×10^{-3}
29a	870	890	0.10
29b	885	913	0.11
29c	891	916	0.12
29d	900	922	0.17

[a] The spectra were taken in chloroform for **28a-c** and in dichloromethane for **29a-d**. Fluorescence quantum yields of **28a-c** relative to methylene blue ($\Phi_f = 0.52$ in chloroform) and **29a-d** relative to indocyanine green ($\Phi_f = 0.13$ in dimethyl sulfoxide).

toluene/*n*-butanol (1:1), show intense absorption at about 700 nm and emission at about 800 nm with rather low quantum yields, respectively (Table 3.2).

Würthner et al. observed an unusual halogen effect in a series of squaraine dyes that can lead to a bathochromic shift of absorption, accompanied by amazing increase of the fluorescence quantum yield. Heavy atoms such as bromine and iodine are commonly deemed to quench the fluorescence because of a perturbation of the fluorescing S_1 state by spin–orbit coupling [76]. They utilized this unusual halogen effect in design of the squaraine-based NIR fluorophores **29a-d** having absorption maxima from 870 nm to 900 nm and the emission maxima up to 922 nm with the quantum yield as high as 17% (Table 3.2) [77]. Starting from squaric acid diethyl ester, condensation reaction with malononitrile afforded the dicyanovinyl-substituted squaric acid derivative, which then reacted with two equivalents of the respective methylene base to afford the desired NIR dyes.

a: $R_1 = R_2 = (CH_3)_2CHCH_2$
b: $R_1 = R_2 = n$-Bu
c: $R_1 = n$-C_3H_7, $R_2 = C_2H_5(CH_3)CH$
d: $R_1 = R_2 = n$-C_3H_7
e: $R_1 = R_2 = C_2H_5$

30

30a, $\lambda_{max}^{abs} = 880$ nm
(toluene/1-butanol, 3/1)

31

31a, $\lambda_{max}^{abs} = 1100$ nm
(toluene/1-butanol, 3/1)

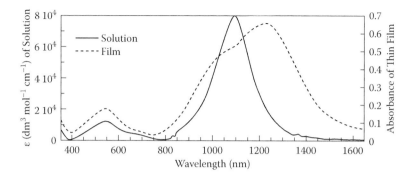

Figure 3.2 Absorption spectra of dye **31a** in 1,2-dichloroethane (solid line) and in a 63-nm-thick, spin-coated film (broken line). (Adapted with permission from M. Tian et al., *J. Am. Chem. Soc.* **2003**, *125*, 348. Copyright (2003) American Chemical Society.)

Croconic acid is considered to be a stronger electron acceptor than squaric acid, which should lead to the chromophores with a relatively lower LUMO level. Tian and coworkers reported the simple synthesis of NIR-absorbing croconates and subsequent manipulation leading to a novel dye with the absorption maximum at 1100 nm in solution [78,79]. Croconate **30a** is readily obtained by condensation of croconic acid with 5-(*N,N*-diisobutylamino)-1,3-benzenediol and shows a maximal absorption band at 880 nm due to the presence of D–π–A–π–D structure, in which the two benzene rings serve as a π-bridging unit, the dialkyl amino groups are donors (D), and the positively charged central ring is an acceptor (A). Further oxidative cyclization of **30** to form the lactone gives compounds **31**. The spin-coated film of **31a** exhibited a broad absorption band with the low-energy peak at 1228 nm. Compared to the absorption of its solution, the absorption of the spin-coated film is red shifted by 135 nm, indicating the formation of J-like aggregates in the solid state.

The origin of NIR absorption of squaraine, croconate, and related dyes has been investigated by high-level calculations. Although they are in general considered to be D–A–D type molecules, calculations have shown conclusively that the NIR absorption in these molecules is due to the diradicaloid nature of the central oxyallyl ring and correlates well to the C–C–C angle of the oxyallyl ring. The one-electron orbital picture shows clearly that smaller HOMO–LUMO gap (HLG) or longer wavelength absorption for croconate derivatives relative to squaraines is due to the larger central C–C–C angle of oxyallyl substructure ($\theta = 87°–92°$ for SQ and 104°–112° for CR, Figure 3.3). The decrease in HLG is due to the destabilization of HOMO and stabilization of LUMO as the C–C–C angle increases, which implies increasing the angle to around 120°, such as rhodizonate (RH) derivatives ($\theta = 115°–123°$), would lead to an even smaller

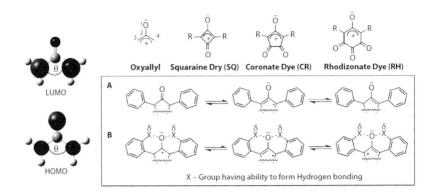

Figure 3.3 Molecular structures of the squaraine (SQ)-, croconate (CR)-, and rhodizonate (RH)-based dyes. Frontier one-electron molecular orbitals for oxyallyl substructure with C–C–C angle dependence. Resonance structures of the oxyallyl subgroups in the presence (B) and absence (A) of hydrogen-bonding groups.

HLG. From the resonance structures of oxyallyl substructure (Figure 3.3), the electron-donating groups, without hydrogen-bonding capability, favor the high diradical character. Increasing the diradical nature a redshift in absorption is seen in non-hydrogen-bonded systems [29]. But in the case of hydrogen-bonded system even with small diradical character, the absorption in the red is achieved. This is due to the larger coupling of diradical and zwitterionic forms [80]. Resonance forms with larger contribution of the zwitterionic part have a longer C = O bond length due to the hydrogen bonding, and a charge separation is retained in the molecule. Accordingly, unusual long absorption of compounds **31** in comparison with **30** can be attributed to the presence of acyclic oxyallyl moiety with a central angle close to 120° and strong intramolecular hydrogen bonding. Furthermore, from the calculations it is predicted that rhodizonate derivatives should absorb at the wavelength above 1000 nm than the corresponding squaraines and croconate dyes. When designing this type of NIR chromophores, more attention should be paid to the use of groups that can perturb the HLG and decrease it without changing the MO character rather than increasing the donor strength [81].

3.3   NIR compounds containing donor–acceptor chromophores

During the past several decades, much effort has been made on the understanding and tuning of the electronic structure of π-conjugated organic compounds through structural manipulation. The use of an electron-donor (high-lying HOMO) and an electron-acceptor (low-lying LUMO) in molecular design is a widely accepted and used approach. The donor and acceptor units can be linked either by a nonconjugated σ spacer or a conjugated π spacer. For the σ-spacer linked compound, since the mixing of HOMO and LUMO orbital is precluded, the HOMO and LUMO level can be tuned independently. Tetrathiafulvalene (TTF) and its derivatives are well known as π-electron donors in molecular conductors. Covalent linkage of TTF or its derivatives and electron acceptors by a σ-spacer yields a charge-transfer molecules that can have a small band gap and are potentially useful as molecular conductor [82], molecular switch [83], and unimolecular rectifier [84]. A variety of compounds with the D–σ–A structure based on TTF or its derivatives and many different acceptors have been extensively studied [85].

One example that is worthwhile noting is the first well-characterized covalently linked TTF–σ–TCNQ (**32**) as reported by Perepichka and coworkers in 2003 [86]. TCNQ (tetracyano-*p*-quinodimethane) is covalently linked to TTF through a nonconjugated 5-atom chain. Cyclic voltammetry (CV) shows two reversible oxidation and the two reversible reduction waves, which are from the TTF and TCNQ moieties, respectively. The difference between the two redox is very small which correlates to a small HOMO–LUMO gap of only 0.17 eV. In solution, the intramolecular charge transfer (ICT) gives rise to a broad absorption band with a maximum at 1630 nm (0.75 eV) and onset at about 2700 nm (0.45 eV), which is assigned to the transition in the head-to-tail conformation (Figure 3.4, left). Similarly, linking TTF with an acceptor through a conjugated spacer gives rise to π-conjugated D–A molecules with a small HOMO–LUMO gap. For example, by fusing TTF with the electron-deficient dipyrido[3,2-*a*:2',3'-*c*] phenazine, D–A compound **33** shows solvent-dependent fluorescence at room temperature (Figure 3.4, right) [87], with a maximal emission at 864 nm in DMF.

34, R_1, R_2 = alkyl, aryl
π = phenylene, 1, 5-thienylene, ethylene, acetylene

35, λ_{max}^{abs} = 584 nm
λ_{max}^{pl} = 775 nm

Among the known π-spacers, thiophene or benzene is often used to connect a donor and an acceptor. With the two donor groups, the D–π–A–π–D chromophores have the enhanced electronic interactions between

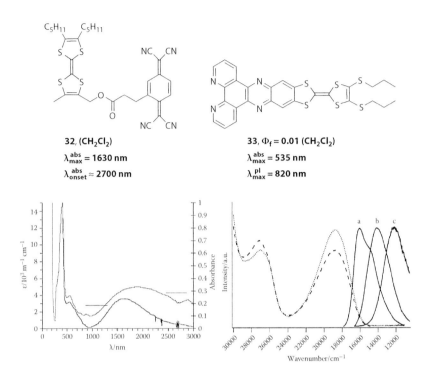

Figure 3.4 (Left) Absorption spectra of compound **32** in CH$_2$Cl$_2$ (lower trace) and in KBr pellet (top trace). (Right) Emission spectra of compound **33** in (a) cyclohexane, (b) toluene, (c) dichloromethane, and absorption (•••••) and excitation spectra (— — -) of **33** in toluene.

donor and acceptor, which can further reduce the HOMO–LUMO gap. For example, a series of D–π–A–π–D type of compounds **34** containing benzo[1,2,5]thiadiazole as an acceptor and *N,N*-diaryl or dialkylamino as a donor that are linked through various conjugated units as a π-spacer were reported [88]. By changing the donors and π-spacers, the absorption and emission colors can be tuned from orange to red but fall short to reach the NIR spectral region. Replacing the sulfur atom in 1,2,5-benzothiadiazole with selenium usually causes a bathochromic shift, but compound **35** still shows absorption and emission maxima in the red and deep red regions at 584 nm and 775 nm, respectively [89].

Table 3.3 Optical properties of compounds **36–38** in toluene

Compound	λ_{max}^{abs} [nm]	Logε	λ_{max}^{pl} [nm]	Φ_f[%]
36a	763	4.38	1065	7.1
36b	837	4.45	1120	2.8
36c	932	4.04	1230	1.8
37a	848	4.71	1055	18.5
37b	954	4.56	1120	4.6
37c	1084	4.48	1285	1.9
38a	920	4.86	1125	5.3
38b	1036	4.48	1295	1.1
38c	1177	4.17	1360	<1

a: X = Y = S
b: X = Se, Y = S
c: X = Y = Se

Wang and coworkers obtained a series of the D–π–A–π–D type of NIR compounds **36–38** by using the strong electron acceptors, namely, benzobis(1,2,5-thiadiazole) (BBTD) and its derivatives [90]. BBTD was firstly reported by Yamashita and deemed to be an efficient electron acceptor for constructing NIR chromophores due to its intrinsic merits [91,92]. One of the merits that the BBTD unit has is the presence of the hypervalent sulfur atom that exhibits a high electron affinity. In addition, BBTD tends to adopt the quinoid form stabilized by low resonance energy. In this series, by changing the donor, acceptor, and π-spacer, the absorption and emission can be tuned within the wavelengths of 600–1400 nm and 900–1600 nm, respectively. A general trend indicates that increasing the strength of donor and acceptor or facilitating the electronic interaction between them can effectively lead to a reduction of the energy gap. For example, a redshift was observed when the fluorene was replaced by the triphenylamine donor (93 nm redshift from **37c** to **38c**). On the acceptor part, replacing the sulfur with selenium led to a large redshift of 116 nm (**38a** versus **38b**) and 141 nm (**38b** versus **38c**). Meanwhile, incorporation of a thiophene bridging unit can result in a further significant bathochromic shift (e.g., 245 nm from **36c** to **38c**), due to the extension of conjugated length and enhanced intramolecular charge transfer. The fluorescence quantum yield decreases as the emission shifts to the longer wavelength, which is due to the increased nonradiative deactivation as the energy gap decreases [93].

To further demonstrate the effect of the acceptor strength on the energy gap, three compounds **39-41** having the same diphenylamino donor and phenylene spacer were synthesized [94]. The optical properties of compounds **39** and **40** are nearly same, indicating that the phenyl groups on the [1,2,5]thiadiazolo[3,4-g]quinoxaline (TQ) moiety affects little the LUMO level. For compound **41**, due to the planar, extended conjugation, its absorption spectrum is redshifted about 70 nm (Figure 3.5). However, compared to those containing the strong BBTD and analogous acceptors (e.g., **36a-c**), compounds **39-41** have a relatively larger energy gap, which shows clearly the dominating role that the acceptor can play in tuning the energy gap of compounds containing the D–π–A–π–D type of chromophores.

The effect of the acceptor strength on the energy gap of the D–π–A–π–D chromophore system is also shown with compounds **42** and **43** having the only EDOT group as a donor [95]. The former is initially used as a monomer to construct low band-gap polymers [96], and the latter is its analog with a weaker acceptor. Compound **43** shows the absorption and emission maxima at 531 nm and 698 nm (Φ_f = 0.21 in CH_2Cl_2), respectively. In comparison, due to the presence of a stronger BBTD acceptor, a significant redshift was observed for compound **42** with the absorption maximum at 650 nm and the emission peak at 805 nm (Φ_f = 0.076 in CH_2Cl_2) (Figure 3.6). Although the increase of acceptor strength can affect the energy gap of the D–A–D type of chromophores, without using a suitable spacer, an ability of tuning the conjugation length and thus energy gaps is rather limited.

44, (CH_2Cl_2)
λ_{max}^{abs} = 700 nm
λ_{max}^{pl} = 1055 nm

45, (CH_2C_2)
λ_{max}^{abs} = 945 nm
λ_{max}^{pl} = 1285 nm

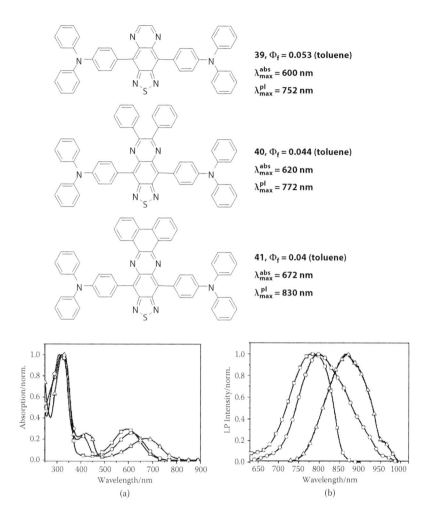

39, $\Phi_f = 0.053$ (toluene)
$\lambda_{max}^{abs} = 600$ nm
$\lambda_{max}^{pl} = 752$ nm

40, $\Phi_f = 0.044$ (toluene)
$\lambda_{max}^{abs} = 620$ nm
$\lambda_{max}^{pl} = 772$ nm

41, $\Phi_f = 0.04$ (toluene)
$\lambda_{max}^{abs} = 672$ nm
$\lambda_{max}^{pl} = 830$ nm

Figure 3.5 (a) Normalized absorption and (b) photoluminescence spectra of compounds **39–41** in dichloromethane, **39** (squares), **40** (circles), and **41** (triangles).

The π-spacer can play an important role in tuning the energy levels of the D–π–A–π–D type of chromophores as it can extend the π-conjugation and promote or stabilize the quinoid structure. Compared to **44**, compound **45** with the same donor and acceptor exhibits a large redshift of 245 nm and the maximal absorption and emission bands at 945 nm and 1285 nm, respectively (Figure 3.7) [97]. The electron-rich thiophene spacer favors the formation of the quinoid structure and facilitates intramolecular charge transfer.

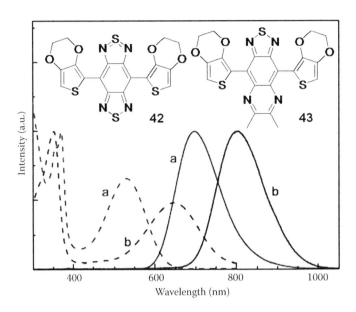

Figure 3.6 Absorption (dashed lines) and photoluminescence (solid lines) spectra of **42** (b) and **43** (a) in dichloromethane.

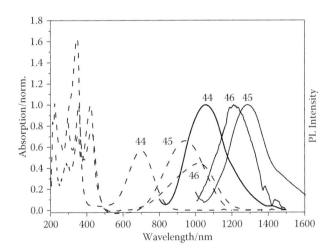

Figure 3.7 Absorption (dashed lines) and photoluminescence (solid lines) spectra of **44, 45,** and **46** in dichloromethane.

46, (toluene)

$\lambda_{max}^{abs} = 1020$ nm

$\lambda_{max}^{pl} = 1213$ nm

47, unknown

$\lambda_{max}^{abs} = 1126$ nm, predicted

$\lambda_{max}^{pl} = 1278$ nm, predicted

By analogy, the use of pyrrole as a spacer should give a further decrease in the energy gap, since pyrrole is more electron-donating than thiophene. Compound **46**, the pyrrole spacer and the BBTD core are expected to be quite coplanar by virtue of intramolecular hydrogen bonding, which should facilitate the charge transfer from donor to acceptor and thus lowers the energy gap. This intramolecular hydrogen bonding also contributes to altering the energy level of the BBTD acceptor by sharing or removing partial electrons on the nitrogen in BBTD, as confirmed by the proton NMR study [98]. Indeed, in comparison with the structural analog **37a**, compound **46** has the absorption peak at 1020 nm and the emission maximum at 1213 nm, representing a large redshift of 172 nm and 158 nm in absorption and emission spectra, respectively. In fact, by comparing the optical properties of the three series of D–π–A–π–D type of chromophores **36–38**, one could quantify the contribution of each of three subunits, namely, donor, acceptor, and π-spacer, to the bathochromic shifts in absorption and fluorescence. For example, the increase of acceptor strength causes relatively large bathochromic shifts of the absorption and fluorescence maxima by 106 nm and 65 nm from **37a** to **37b**, respectively, and by 130 nm and 165 nm from **37b** to **37c**, respectively. The same trend can be seen for the other two series of chromophores **36a-c** and **38a-c**. In comparison, the increase of donor strength from the fluorene group to triphenylamino group can bring about redshifts of 70–93 nm in absorption and 72–175 nm in fluorescence. Finally, change of the π-spacer from benzene to thiophene has a profound effect on bathochromic shift, as evident by 245-nm shift in absorption (**36c** to **38c**) and 175-nm shift in fluorescence (**36b** to **38b**). By applying these subunit contributors to compound **46**, one could predict the optical properties of its unknown analogs **47**. With increase of acceptor strength from **46** to **47**, as in a similar case of **37a** to **37b**, unknown compound **47** would have the absorption and fluorescence maxima at 1126 nm and 1278 nm, respectively.

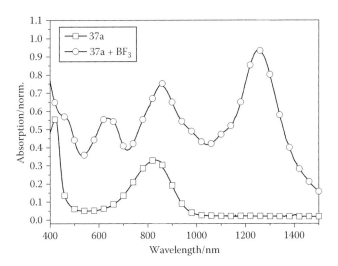

The effect of the intramolecular hydrogen bonding on the bathochromic shift also implies that by removing the electrons from the acceptor one can expect a further decrease in energy gap or a spectroscopic redshift. This rational predication has recently been demonstrated by adding Lewis acid (e.g., BF_3) to a benzothiadiazole compound [99]. Figure 3.8 shows the absorption spectra of compound **37a** and a mixture of **37a** with excess BF_3. Upon addition of an excess of BF_3, a color change took place immediately, going from yellowish green to dark blue visually. Consequently, the λ_{max} and λ_{onset} of compound **37a**-BF_3 adduct appeared at 1260 nm and 1580 nm, being red-shifted by 426 nm and 622 nm, respectively [98].

In addition to increasing the acceptor strength by making it more electron deficient, lowering the LUMO level of the acceptor unit in the

Figure 3.8 Normalized absorption spectra of compound **37a** (squares) and a mixture of **37a** with an excess of BF_3 (circles). (Adapted with permission from G. Qian, Z. Y. Wang, *Can. J. Chem.* **2010**, *88*, 192.)

D–π–A–π–D chromophore system by extending its conjugation is another efficient way to narrow the energy gap. Many electron-deficient heterocyclics are known, but none of the D–A chromophores based on them as an acceptor are NIR-absorbing, such as hexaazatriphenylene (HAT) [100–104] and quinoxalino[2′,3′,9,10]phenanthro [4,5-*abc*]phenazines (PPP) [105,106]. The tetraphenyl-substituted PPP (PPP-Ph) and HAT (HAT-Ph) were reported to have the energy gaps of 2.78 and 2.85 eV, or HOMO/LUMO energies of 5.92/3.25 and 6.37/3.70 eV, respectively [105,106]. The calculations indicate that PPP-Ph and hexaphenyl-substituted HAT (HAT-Ph) have a similar energy gap of 3.30 and 3.20 eV, respectively. Theoretical calculations show that by fusing thiadiazole units onto PPP and HAT cores, the LUMO levels decrease from 2.70 and 3.10 eV to 3.50 and 3.80 eV, respectively, resulting in a significant decrease in the energy gap for hypothetic PPP-BT (2.30 eV) and HAT-BT (2.20 eV) [107].

48a , λ^{abs}_{max} = 932 nm

λ^{pl}_{max} = 1250 nm

48b , λ^{abs}_{max} = 1003 nm

λ^{pl}_{max} = 1290 nm

49a, λ^{abs}_{max} = 746 nm

λ^{pl}_{max} = 1035 nm

49b, λ^{abs}_{max} = 844 nm

λ^{pl}_{max} = 1100 nm

50: $\lambda_{max}^{abs} = 530$ nm

$\lambda_{max}^{pl} = 700$ nm

Based on these computational study and predication, a new series of low energy gap chromophores (**48** and **49**) were designed, synthesized, and characterized. Compounds **48** and **49** have the energy gap of 1.27–0.71 eV, and accordingly absorb at 746–1003 nm and emit at 1035–1290 nm in toluene. In comparison with compound **50** containing the same donor units [108], there is a significant bathochromic shift with 402 nm and 550 nm for absorption and emission for compound **48a**, respectively. Figure 3.9 shows the calculated LUMOs of the reference compounds along with the synthesized compounds. When comparing PPP-BT versus PPP-Ph, it is clear that the benzothiadiazole moiety of the former compound gives an additional delocalization advantage. Furthermore, the steric effect on the effective conjugation and the energy gap is clearly shown, as the two neighboring phenyl groups in PPP-Ph are highly twisted and those in PPP-BT are less twisted relative to the core. Accordingly, this design feature allows for a better delocalization or more efficient charge transfer, as evident by the delocalization extended to the electron donor moieties of **49a** and **49b**, in particular, even to the fluorene moiety of **49b**. When comparing the hexaazatriphenylene series, it shows again that the steric interference of vicinal phenyl groups in HAT-Ph does not allow perfect delocalization. The delocalization in compounds **48a** and **48b** extends further into the side chains, creating large chromophores with smallest gaps (0.91 and 0.71 eV, respectively).

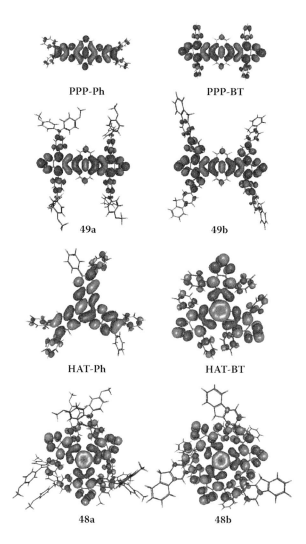

Figure 3.9 (See color insert.) Comparison of LUMO spatial distribution (isovalue = 0.01 a.u) between reference chromophores PPP-Ph, PPP-BT, HAT-Ph, and HAT-BT and synthesized chromophores **48** and **49**.

3.4 NIR compounds containing metal complex chromophores

Small HOMO–LUMO gaps are known to result from some metal-to-ligand or ligand-to-metal charge transfer (MLCT or LMCT) process [109], where the energy barrier for the ground state electron transfer between metal and ligand is high enough to prevent an intramolecular redox process. Typically, the two types of metal complex chromophores, namely, the metal complexes containing radical ligands with the MLCT and LMCT and the mixed-valence di- or tri-nuclear metal complexes capable of having the metal-to-metal charge transfer (MMCT), are able to display strong NIR absorption. It should be pointed out that the most NIR-absorbing metal complexes are usually generated in situ during the electrochemical redox and may not be stable in air or difficult to purify. This particular property may hamper the direct use of some NIR-absorbing metal complexes, especially those containing highly reactive radical species, for certain applications but are suitable for use in an electrochemical device, in which the electrochemical redox reaction can proceed.

Most of neutral ligand molecules in their ground states are characterized by large HOMO–LUMO gaps, thus displaying the absorption bands in the UV and visible spectral region. One-electron oxidation or reduction generates singly occupied molecular orbitals (SOMOs) for the corresponding radical cations and radical anions, offering a possibility for low-energy transitions and often for NIR absorption. The propensity for low-energy absorptions by radical ions has been noted, and the corresponding electronic absorption data have been well documented [110].

Early work by Lever et al. has demonstrated that ligand-based redox processes between the catecholate (cat), semiquinonate (sq), and quinone (q) forms are responsible for the different oxidation states, while the metal center remains in the +2 oxidation state throughout [111]. The complexes display a strong NIR absorption in their semiquinone forms due to a $Ru(d\pi) \rightarrow sq(\pi^*)$ MLCT band that is quenched both upon oxidation (to the quinone form) and upon reduction (to the catechol form) [112]. Schwab et al. reported a series of $Ru(bpy)_2$-dioxolene complexes (**51–54**), along with analogous compounds **55–58,** where 2,2'-bipyridine-4,4'-dicarboxylic acid ligands are replaced by bpy ligands [113].

51: M = M_1; 55: M = M_2 53: M = M_1; 57: M = M_2

52: M = M_1; 56: M = M_2 54: M = M_1; 58: M = M_2

M_1: R = H
M_2: R = COOH

Ru complexes	51	52	52	54	55	56	57	58
MLCT band, λ_{max} (nm)	876	974	948	942	892	978	953	945

All the Ru complexes display the expected Ru(dπ) → sq(π*) MLCT band that can be activated and deactivated reversibly by simply switching between a positive and negative bias very close to 0 V versus SCE. The spectroelectrochemical studies indicate that the maximal absorption is a function of the dioxolene ligand and ranges from 880 to about 974 nm (Figure 3.10).

Peter and Ward et al. reported a similar work at almost the same time [114], involving the use of [Ru(dcbpy)$_2$(Cl$_4$cat] (dcbpy = 2,2′-bipyridine-4,4′-dicarboxylic acid) for modification of the surface of nanocrystalline antimony-doped tin oxide (SnO$_2$:Sb) electrodes. The spectroelectrochemical study in water shows an absorption peak at 940 nm arising from a Ru→semiquinone MLCT transition upon oxidation of the dioxolene ligand, similar to those observed for the noncarboxylated parent complex [Ru(bpy)$_2$(Cl$_4$cat)] [111]. Due to their very interesting spectroelectrochemical behavior, a variety of 1,2-dioxolene-containing Ru complexes have been investigated in depth [115].

Mixed-valence compounds contain an element which, at least in a formal sense, exists in more than one oxidation state [116]. The mixed-valence character of some minerals provides the basis for their color. Prussian blue {Fe^{III}$_4$[Fe^{II}(CN)$_6$]3} has a cyanide-bridged Fe(II)-Fe(III) structure and is one of the first metal-valence compounds [117]. Multiple-site metalloenzymes, which undergo multiple electron transfer, exist in the mixed-valence forms. Mixed-valence materials were classified earlier by Robin and Day [118] and reviewed by Allen and Hush [119]. In the 1970s, Creutz

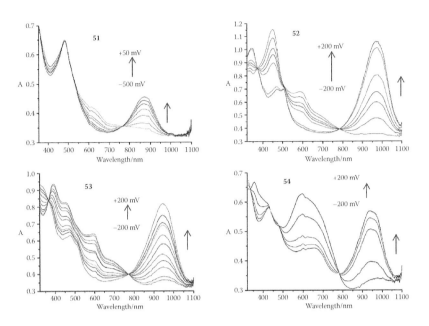

Figure 3.10 Absorption spectral changes of the Ru(bpy)$_2$ dioxolene complexes **51–54** in acetonitrile solution. (Adapted with permission from P. F. H. Schwab et al., *Inorg. Chem.* **2003**, *42*, 6613. Copyright (2003) American Chemical Society.)

and Taube designed and prepared the first mixed-valence complexes, $[(NH_3)_5Ru(pyrazine)Ru(NH_3)_5]^{5+}$ [120]. The understanding of the mixed-valence metal complexes further advanced in the field of coordination chemistry by Crutchley [121], and Meyer et al. [122]. More recently, the scope of the mixed-valence metal complexes has been expanded by Kaim et al. [123]. Using the Creutz–Taube ion as a benchmark, the extensive studies have shown that replacement of the neutral donor of ammonium by the anionic cyanide in $[(NC)_5Ru(\mu\text{-pyrazine})Ru(CN)_5]^{5+}$ causes a decreased electrochemical stability and a broadened NIR absorption from the intervalence charge transfer (IVCT) [124]. Subsequent change of the metal from ruthenium to iron in $[(NC)_5Fe(\mu\text{-pyrazine})Fe(CN)_5]^{5+}$ results in a significant bathochromic shift, due to weakened metal–metal interaction [125]. The $[(NC)_5Fe(\mu\text{-pyrazine})Fe(CN)_5]^{5+}$ ion exhibits two bands at 599 and 745 nm (Figure 3.11), which are assigned as split MLCT components. The band at about 400 nm could also contain a cyanide to iron(III) LMCT component. In the mixed-valence states, the expected IVCT transitions appear in the NIR region with a maximum at 2475 nm. Furthermore, the pyrazine-bridged metal complexes having Mo and W can also show intense NIR bands around 2100 nm during the electrochemical redox process [126].

To be useful in optoelectronic devices, mixed-valence metal complexes must fulfill a number of key requirements, at least including availability

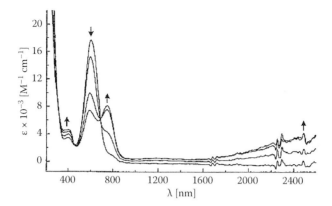

Figure 3.11 UV–Vis–NIR absorption spectra of [(NC)₅Fe(μ-pyrazine)Fe(CN)₅]⁶⁺ during the electrochemical oxidation using OTTLE in acetonitrile (0.1 M Bu₄NPF₆). (Adapted with permission from M. Ketterle et al., *Inorg. Chim. Acta,* **1999,** *291,* 66.)

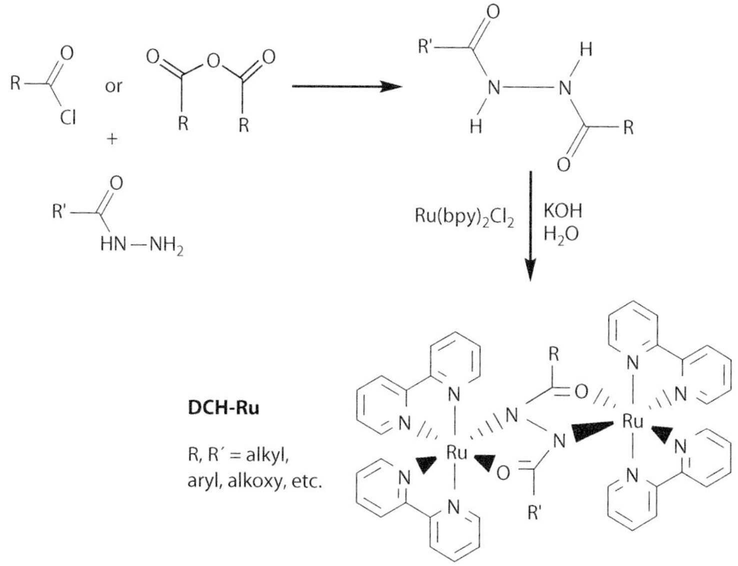

Scheme 3.2 General synthesis of **DCH-Ru** complexes.

in large quantity, good stability in air, good processability, and excellent electrochemical property. The early work by Kaim et al. shows that dinuclear ruthenium complexes (**DCH-Ru**, Scheme 3.2) with the bridging unit of 1,2-dicarbonylhydrazido (dch) and the ligand of 2,2′-bipyridine (bpy) all have the MMCT for the mixed-valence state in a range of 1000 to 1800 nm [127]. From Kaim's work, these complexes, typically isolated in the

Ru(III)–Ru(II) oxidation state, possess two nondegenerate Ru(II) → Ru(III) redox couples. As well, in addition to the MMCT transition the complexes possess strong MLCT and LMCT associated with the Ru(II)/Ru(II) and Ru(III)/Ru(III) oxidation states, respectively. The complex of [{Ru(bpy)$_2$}$_2\mu$-(dch-OC$_2$H$_5$)]$^{n+}$, for instance, possesses strong bipyridyl centered MLCT (d Ru(II) → π^* bpy) around 520 nm, LMCT [π (dch-OC$_2$C$_5$)$^{2-}$ → d Ru(III)] at 900 nm, and MMCT [d Ru(II) → d Ru(III)] at 1420 nm.

DCH-Ru complexes can be readily synthesized in good yields (typically above 70%) from a variety of dicarbonylhydrazines with Ru(bpy)$_2$Cl$_2$ in alkaline water in open air (Scheme 3.2). These complexes are chemically stable in air at elevated temperatures (e.g., 200°C). The complex can be considered analogous to a fused ring system, and as such, a substituent influence could selectively perturb the metal center associated with the ring bearing the particular substituents. A series of symmetrical and unsymmetrical **DCH-Ru** complexes were prepared for gaining a better understanding of correlation between the molecular structure and both electrochemical and optical properties [128].

All the **DCH-Ru** complexes exhibited two quasi-reversible positive one electron redox couples associated with the ruthenium metal centers and, in some complexes only, two reversible, negative two-electron couples associated with the bipyridine ligands. As the average donor strength of the substituent increases, there is a shift of both the first ($^1E_{1/2}$) and second oxidation potentials ($^2E_{1/2}$) to lower values (Table 3.4). This can be attributed to the increase of the stability of ruthenium d-orbitals when

Table 3.4 Electrochemical data
for **DCH-Ru** complexes

R, R′	$^1E_{1/2}$	$^2E_{1/2}$	ΔE
Ph, N(CH$_3$)$_2$	650	1150	500
Ph, CH$_3$	730	1330	600
Ph, CH$_3$OPh	790	1360	570
Ph, Ph	800	1380	580
Ph, NO$_2$Ph	880	1480	600
Ph, CF$_3$	970	1540	570
CH$_3$OPh, N(CH$_3$)$_2$	630	1080	450
CH$_3$OPh, CH$_3$	740	1320	580
CH$_3$OPh, CH$_3$OPh	760	1330	570
CH$_3$OPh, NO$_2$Ph	850	1420	570
CH$_3$OPh, CF$_3$	960	1520	560
NO$_2$Ph, NO$_2$Ph	930	1510	580

Note: From cyclic voltammetry performed at 200 mV/s scan rate. Potentials E in mV versus NHE.

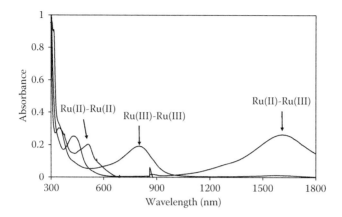

Figure 3.12 UV–vis–NIR spectra of [{Ru(bpy)$_2$}$_2$$\mu$-NO$_2$ph,ph-dch]$^{n+}$ in acetonitrile (0.1 M TBAH) using the OTTLE cell.

the ligand has poor Hammett donor properties. Since the influence for the two metal centers are nearly equal, the difference ΔE between two oxidation potentials are not significantly affected, where the magnitude of ΔE indicates the stability of the mixed-valance species. The only effect on ΔE is a slight drop over the series with increasing donor strength. By pairing an electron-donating substituent on one ring with that of an electron-withdrawing substituent on the other ring, one ruthenium could be shifted cathodically while the other shifted anodically. Therefore, the effect would be to widen the gap between the individual metal Ru(II) → Ru(III) redox couples, which is important to selectively control an electro-chemical switching device at one particular applied voltage.

As a representative, the absorption spectra of [{Ru(bpy)$_2$}$_2$$\mu$-NO$_2$Ph,Ph-DCH]$^{n+}$ in different oxidation states are shown in Figure 3.12. For the Ru(II)/Ru(II) and fully oxidized Ru(III)/Ru(III) states, they show the bands at 540 nm and 800 nm, due to the MLCT and LMCT transitions, respectively. Only oxidation of DCH-Ru complexes to the mixed-valence state Ru(II)/Ru(III) results in the appearance of an intense NIR band between 1300–1800 nm, which is assigned to the MMCT transition. There are hardly solvatochromic effects for these MMCT bands, indicating a delocalized situation in the mixed-valence formulation.

Over the series, the MMCT band is blueshifted as the donor potential of substituent increases (Table 3.5). This can be rationalized in a hole-transfer scheme, which involves a concerted electron transfer from the bridge to the acceptor metal and an electron transfer from the donor metal to the bridge. The overall effect of the increase in electron density in the bridge is to destabilize the Ru(II) donor orbitals while stabilizing the Ru(III) acceptor orbitals, which has the effect of increasing the energy

Table 3.5 Electronic spectral data for **DCH-Ru** complexes in acetonitrile

R	n+	$\lambda_{max}(\log \varepsilon)$
Ph	2	245(4.96), 294(4.92), 356(4.29), 518(4.19)
	3	243(4.72), 292(4.88), 432(4.10), 1590(4.13)
	4	250(4.80), 306(4.65), 317(4.66), 808(4.14)
NO$_2$Ph	2	288(4.69), 348(4.25), 507(4.19)
	3	288(4.63), 430(3.87), 1629(3.83)
	4	305(4.43), 315(4.42), 779(3.76)
CH$_3$OPh	2	290(4.67), 354(4.24), 520(4.13)
	3	289(4.65), 440(3.86), 1593(3.84)
	4	303(4.47), 850(3.87)
NH(CH$_2$CH$_2$CH$_3$)	2	288(5.01), 454(4.17)
	3	287(5.03), 451(4.18), 685(3.63), 795(3.60), 1119(3.43)
	4	286(4.97), 441(4.11), 639(3.72), 792(3.71)
	5	305(4.81), 642(3.37)
Ph, CF$_3$	2	244(4.67), 353(4.28), 498(4.21)
	3	244(4.68), 287(4.86), 425(4.08), 1655(4.04)
	4	244(4.73), 304(4.77), 311(4.77)
Ph, NO$_2$Ph	2	245(4.66), 293(4.91), 352(4.25), 513(4.18)
	3	245(4.68), 290(4.84), 429(4.05), 1612(4.07)
	4	249(4.76), 304(4.63), 801(3.93)
Ph, CH$_3$OPh	2	244(4.69), 294(4.88), 354(4.25), 519(4.16)
	3	244(4.70), 291(4.84), 434(4.08), 1590(4.07)
	4	247(4.72), 284(4.65), 824(3.97)
Ph, CH$_3$	2	245(4.65), 292(4.93), 350(4.24), 522(4.18)
	3	243(4.65), 291(4.88), 431(4.03), 1557(4.08)
	4	248(4.74), 306(4.60), 316(4.61), 812(4.12)
Ph, N(CH$_3$)$_2$	2	245(4.67), 354(4.23), 519(4.13)
	3	244(4.66), 441(4.08), 1378(3.77)
	4	245(4.68), 281(4.80), 911(4.02)
Ph, NH(CH$_2$CH$_2$CH$_3$)	2	246(4.66), 348(4.20), 521(4.12)
	3	244(4.66), 291(4.90), 429(4.02), 1246(3.71)
	4	248(4.70), 281(4.74), 363(4.01), 631(3.55), 943(4.01)
CH$_3$OPh, CF$_3$	2	242(4.67), 354(4.24), 499(4.16)
	3	245(4.67), 428(4.04), 1639(3.91)
	4	251(4.78), 304(4.74), 311(4.74)
CH$_3$OPh, NO$_2$Ph	2	244(4.67), 288(4.93), 458(4.06)
	3	243(4.59), 285(4.79), 1600(3.54)
	4	246(4.52), 283(4.55), 319(4.44), 805(3.78)

Table 3.5 (continued) Electronic spectral data for **DCH-Ru**
complexes in acetonitrile

R	n+	$\lambda_{max}(\log \varepsilon)$
CH$_3$OPh, CH$_3$	2	245(4.69), 348(4.23), 523(4.15)
	3	244(4.78), 446(4.10), 1556(4.04)
	4	251(4.75), 304(4.65), 820(4.12)
CH$_3$OPh, N(CH$_3$)$_2$	2	244(4.66), 288(4.95), 351(4.16)
	3	244(4.65), 288(4.95), 450(4.14), 1354(3.58)
	4	245(4.62), 285(4.92), 387(3.98), 419(3.97), 909(3.92)
CH$_3$OPh,	2	245(4.65), 289(4.95), 346(4.15), 455(4.10)
NH(CH$_2$CH$_2$CH$_3$)	3	244(4.65), 288(4.94), 449(4.10), 1253(3.57)
	4	246(4.65), 285(4.88), 929(3.91)
NO$_2$Ph,	2	245(4.65), 288(4.97), 455(4.13)
NH(CH$_2$CH$_2$CH$_3$)	3	244(4.62), 286(4.94), 427(4.10), 1224(3.05)
	4	244(4.64), 284(4.86), 425(4.03), 907(2.96)

Note: Wavelengths in nm and molar extinction coefficients ε in M^{-1}cm^{-1}.

gap between donor and acceptor metals. The structure–property correlation studies on **DCH-Ru** complexes clearly show that the NIR absorption
depends on the electronic environment of the two metal cations, which is
strongly influenced by the substituents on the ligands.

Further to linear **DCH-Ru** complexes, trimeric analogs **59** and **60**
having reactive hydroxy groups were synthesized, for the purpose of
increasing the optical density and film-forming ability (Figure 3.13) [129].
As expected, the trimer complexes displays excellent electrical/optical
properties including a strong NIR band with the maximum absorption
at 1549 nm, large comproportionation constant (K_c = 5.5 × 10^{10}) and com-
paratively low first oxidation potentials at 817 mV versus NHE. The trimer
complexes in the NIR-absorbing state appear to be light yellow in solution
and are quite stable in air. The dendritic structure of trimer complexes
has a large free volume, which allows for rapid ion transportation during
an electrochemical redox process. Compared with linear **DCH-Ru** complexes, complexes **59** and **60** contain more ruthenium units per molecule,
giving rise to a slightly higher molar extinction coefficient of MMCT band
(1.4 × 10^4 M^{-1}cm^{-1}), and also is capable of reacting with polyisocyanates
and forming a film on an electrode.

Regarding the bridging ligand, the basic complex-forming unit in
dicarbonylhydrazines is the CO-N-N-CO moiety. In comparison, the
oxamides contain the N-CO-CO-N moiety, which is a structural isomer
to the one in dicarbonylhydrazines and then should also be qualified as
a bridging ligand. A series of dinuclear ruthenium complexes (**OXA-Ru**)

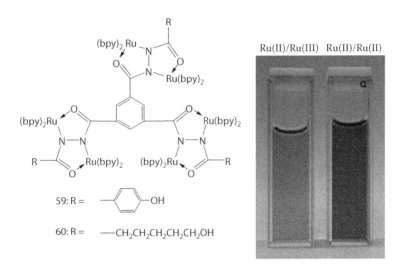

Figure 3.13 (See color insert.) DCH-Ru trimers and their solutions in the Ru(II)/Ru(II) and Ru(II)/Ru(III) states.

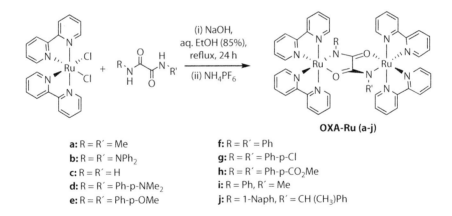

OXA-Ru (a-j)

a: R = R′ = Me

b: R = R′ = NPh$_2$

c: R = R′ = H

d: R = R′ = Ph-p-NMe$_2$

e: R = R′ = Ph-p-OMe

f: R = R′ = Ph

g: R = R′ = Ph-p-Cl

h: R = R′ = Ph-p-CO$_2$Me

i: R = Ph, R′ = Me

j: R = 1-Naph, R′ = CH (CH$_3$)Ph

Scheme 3.3 General synthesis of **OXA-Ru** complexes.

as analogs to **DCH-Ru** complexes were prepared (Scheme 3.3) [130]. The purple-colored **OXA-Ru** complexes were isolated upon addition of ammonium hexafluorophosphate and then passed through an alumina-gel column (eluted with a mixture of acetonitrile and toluene, 1:1 v/v), which allowed for the isolation of pure complexes in the Ru(II)/Ru(II) state.

The spectroelectrochemical studies indicated that **OXA-Ru** have similar electrochromic properties as **DCH-Ru**. As shown in Figure 3.14 for one of **OXA-Ru**(f), two intense absorptions around 370 and 506 nm

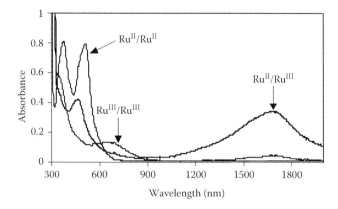

Figure 3.14 UV/vis–NIR spectra of **OXA-Ru**(f) in three different oxidation states in acetonitrile containing 0.1 M TBAPF$_6$.

are associated with the MLCT transitions. Oxidation to the mixed-valence state results in the appearance of a strong and broad band centered in the NIR region. Complex **OXA-Ru**(b) had the most blueshifted NIR absorption with the λ_{max} at 1393 nm, followed by **OXA-Ru**(d) at 1474 nm. The rest of **OXA-Ru** exhibited λ_{max} values around 1650 nm. Upon further oxidation to the Ru(III)/Ru(III) state, the intense NIR absorptions disappear and broad peaks appear around 700 nm. Interestingly, **OXA-Ru**(c), with no N-substitution in the bridging ligand had only one redox couple and did not show any NIR absorption upon oxidation. A possible reason could be due to difficulty in the formation of a stable mixed-valence state, or the amide moiety being more easily oxidized than the Ru metal ion.

3.5   Selective NIR-absorbing compounds

Research on low band-gap organic compounds and polymers has so far mainly focused on achieving bathochromic shifts of the absorption maximum by altering the chromophoric system. Nearly all the known NIR-absorbing compounds and polymers are also highly colored. A question of the existence of colorless, NIR-absorbing organic materials is not only scientifically intriguing but also practically useful in the future of technological innovation. Colorless or barely colored, NIR-absorbing materials can be envisioned for use in a wide range of applications, such as security printing (bank notes, credit cards, identity cards, passports, lottery tickets, etc.), invisible and NIR readable bar codes, laser welding of plastics, infrared radiation absorbers for plasma display panels, laser marking, infrared absorbers for heat management, and so on.

Inorganic materials such as lanthanum hexaboride, indium tin oxide (ITO), antimony tin oxide (ATO) in nanoparticle form, and coated mica materials (Lazerflair® from Merck) are typical examples of colorless NIR absorbers. Several classes of organic dyes show a potential as colorless NIR absorbers, including cyanines, squaraines, croconaines, phthalocyanines and naphthalocyanines, dithiolene, and other metal complexes. However, some of these dyes are chemically instable and some still show rather intense 01 and 02 vibronic bands (Franck–Condon principle) due to the relatively large changes in bond length accompanying the $S_0 \rightarrow S_1$ transition. Thus, their $S_0 \rightarrow S_1$ absorption is extended into the visible range.

61 **62**

R	X	Y	λ_{max} (nm)	$\varepsilon \times 10^{-4}$
4-Me	S	S	890	1.40
H	NH	S	894	1.20
H	NH	Se	858	1.42
4-Me	NH	NH	790	5.32

M	R	λ_{max} (nm) ($\varepsilon \times 10^{-4}$)
Ni	i-Pr	1001 (7.90)
Pd	i-Pr	1020
Pt	i-Pr	1000 (11.3)

Early work in the 1980s reveals that a series of nickel complexes **61** obtained from phenylenediamines and related S and Se analogous ligands show fairly strong NIR absorption and weak visible color [131]. The NIR absorption maxima appear at the wavelengths in a range of 800–900 nm. A patent discloses the invention of some dithiolene metal complexes **62** for use as colorless infrared absorbers, having a generic structure where M is Ni, Pd, Pt, or other metal ions [132]. These metal complexes show absorption maxima typically around 1000 nm regardless of the metal present in the complexes, with one exception for the Pt dithiolene complex that shows a noticeably high absorption coefficient. The inventors claimed that a printing ink formulated using a small amount of one of the NIR-absorbing complexes afforded the visually colorless print that was clearly visible using an IR-viewing device with a 715-nm cutoff filter and showed the excellent fastness to light.

Using 1,2,4,5-benzentetrathiolate as a connecting ligand, Donahue et al. developed a facile route to the synthesis of multimetal dithiolene complexes, such as **63a,b** (Figure 3.15). The central metal ion can affect the overall optical property in terms of the peak position and shape. These two complexes show relatively low absorbance in the visible spectral range between 450 and 750 nm and thus appear to be fairly less colored.

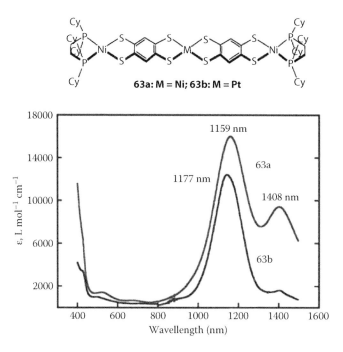

63a: M = Ni; 63b: M = Pt

Figure 3.15 Absorption spectra in DMF solution for **63a** and **63b**. (Adapted with permission from K. Arumugam et al., *Inorg. Chem.* **2009**, *48*, 10591. Copyright (2009) American Chemical Society.)

In comparison with analogous **61** and **62**, linking of the dithiolene-M units to form oligomers **63** results in a noticeable bathochromic shift of 100–200 nm, which implies that further extension of chain length or making a high polymer of **63** could lead to the NIR absorption over 1200 nm.

The Müllen group has been studying and developing NIR-absorbing polyaromatic dyes since early 1990s [12,134]. By extending the π system of perylene diimide **1** along the molecular long axis, they obtained the higher homologues terrylene diimide **2** and quaterrylene diimide **3**. In comparison to the absorption maximum of perylene diimide, those of **2** and **3** are shifted bathochromically (Figure 3.16). Although quaterrylene diimide **3** absorbs strongly in the NIR region (λmax = 762 nm), it is still deeply colored [10]. By further extension of the π-system, they found some interesting optical properties associated with pentarylene diimide **5** and hexarylene diimide **6** [11]. With an increasing degree of annulation, the absorption maximum of **5** and **6** lies at 877 nm and 953 nm, respectively. Not only does a bathochromic shift become apparent but the absorption coefficients increase also phenomenally: ε = 235 M⁻¹ cm⁻¹ for **5** and ε = 293 M⁻¹ cm⁻¹ for **6**. The dilute solutions of **5** and **6** are almost colorless, as their absorption spectra only display small peaks between 650–750 nm

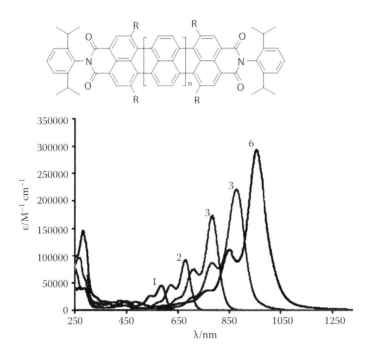

Figure 3.16 Absorption spectra of rylene diimides in chloroform: perylene diimide **1** ($n = 0$, R = H, $\lambda^{abs}_{max} \approx 590$ nm), terrylene diimide **2** ($n = 1$, R = H, $\lambda^{abs}_{max} = 664$ nm), quaterrylene diimide **3** ($n = 2$, R = H, $\lambda^{abs}_{max} = 762$ nm), pentarylene diimide **5** ($n = 3$, R = 4-*t*-octylphenoxy, $\lambda^{abs}_{max} = 877$ nm), and hexarylene diimide **6** ($n = 4$, R = 4-*t*-octylphenoxy, $\lambda^{abs}_{max} = 953$ nm). (Adapted with permission from N. G. Pschirer et al., *Angew. Chem. Int. Ed.* **2006**, *45*, 1401.)

(Figure 3.16), and exhibit an impressive photostability over weeks in sunlight. A negligible absorption in the visible region together with the extraordinary high extinction coefficient at the wavelength of fundamental laser beam (1064 nm) makes rylene dye **6** ideal for some potential applications.

In 2011, Zumbusch et al. reported a series of highly NIR-selective colorless bis(pyrrolopyrrole) cyanine dyes, as represented by **64** [135]. They are extended chromophores based on diketopyrrolopyrroles and characterized by extremely high extinction coefficients (e.g., 556,000 M^{-1} cm^{-1}), narrowband absorption in the NIR around 900 nm (Figure 3.17). The design strategy evolves on the basis of understanding of the relationship between chemical structure and optical properties of rylene dyes and focuses on diminishing the bond length alternation, decreasing the e_{01}/e_{00}-values, and stiffing the extended linear π system. Accordingly, by complexation with boron, all the 11 aromatic rings are locked in place in **64**, leading to a sharpening of the vibronic bands, an increase in the

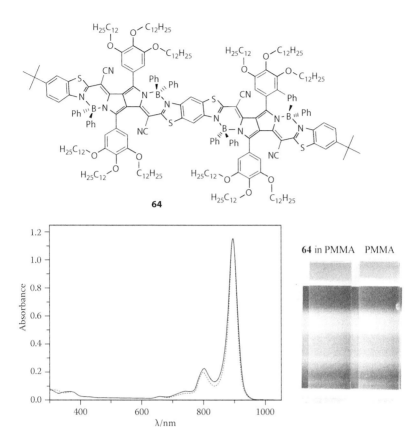

64

Figure 3.17 (See color insert.) Absorption spectra of **64** in chloroform (dotted line) and in PMMA (solid line). A photograph of the PMMA-cuboid containing **64** (left) and the PMMA-cuboid without dye (right). (Adapted with permission from G. M. Fischer et al., *Angew. Chem. Int. Ed.* **2011**, *50*, 1406.)

absorption coefficients, and a shift of the Franck–Condon factors in favor of the 00 transition. In essence, compound **64**, having the electrons being delocalized over a long quasi-linear ladder-like system, acts as one chromophoric molecule. This strategy may be generally applicable to the design and synthesis of many selective NIR-absorbing colorless oligomers and polymers.

Since the visible absorption is negligible, compound **64** is close to an ideal colorless NIR absorber. It can be incorporated in poly(methyl methacrylate) (PMMA), which gives a colorless polymer with the absorption almost identical to that of a solution in chloroform and good transparency as indicated by the background color plate in comparison with the pure PPMA sample (Figure 3.17). The absorption of dye-doped polymer remained almost unchanged after half a year of exposure to daylight. The striking NIR properties and good photostability of this class of colorless dyes make them another promising candidate, such as rylene dyes, for various technical innovations.

References

1. C. R. Fincher, M. Ozaki, M. Tanaka, D. Peebles, L. Lauchlan, A. J. Heeger, A. G. MacDiarmid, *Phys. Rev. B*. **1979**, *20*, 1589.
2. T. Katakami, K. Fukui, T. Okamoto, M. Nakagawa, *Bull. Chem. Soc. Jpn.* **1976**, *49*, 297.
3. E. Clar, *Chem. Ber.* **1948**, *81*, 52.
4. A. Bohnen, K.-H. Koch, W. Lüttke, K. Müllen, *Angew. Chem. Int. Ed.* **1990**, *29*, 525.
5. E. Z. M. Ebeid, S. A. El-Daly, H. Langhals, *J. Phys. Chem.* **1988**, *92*, 4565.
6. W. Herbst, K. Hunger, *Industrial Organic Pigments, 2nd ed.*, VCH: Weinheim, 1997.
7. M. Sadrai, L. Hadel, R. R. Sauers, S. Husain, K. Krogh-Jespersen, J. D. Westbrook, G. R. Bird, *J. Phys. Chem.* **1992**, *96*, 7988.
8. P. Panayotatos, D. Parikh, R. Sauers, G. Bird, A. Piechowski, S. Husain, *Solar Cells*, **1986**, *18*, 71.
9. F. Nolde, J. Qu, C. Kohl, N. G. Pschirer, E. Reuther, K. Müllen, *Chem. Eur. J.* **2005**, *11*, 3959.
10. H. Quante, K. Müllen, *Angew. Chem. Int. Ed.* **1995**, *34*, 1323.
11. N. G. Pschirer, C. Kohl, F. Nolde, J. Qu, K. Müllen, *Angew. Chem. Int. Ed.* **2006**, *45*, 1401.
12. C. Kohl, S. Becker, K. Müllen, *Chem. Commun.* **2002**, 2778.
13. H. Langhals, P. Blanke, *Dyes Pigm.* **2003**, *59*, 109.
14. (a) C. L. Eversloh, C. Li, K. Müllen, *Org. Lett.* **2011**, *13*, 4148. (b) Y. Li, L. Xu, T. Liu, Y. Yu, H. Liu, Y. Li, D. Zhu, *Org. Lett.* **2011**, *13*, 5692.
15. (a) H. Qian, C. Liu, Z. Wang, D.Zhu, *Chem. Commun.* **2006**, *44*, 4587. (b) H. Choi, S. Paek, J. Song, C. Kim, N. Cho, J. Ko, *Chem. Commun.* **2011**, *47*, 5509. (c) W. Jiang, Y. Li, W. Yue, Y. Zhen, J. Qu, Z. Wang, *Org. Lett.* **2010**, *12*, 228. (d) Y. Li, Y. Li, J. Li, C. Li, X. Liu, M. Yuan, H. Liu, S. Wang, *Chem. Eur. J.* **2006**, *12*, 8378.
16. (a) U. Rohr, P. Schlichting, A. Böhm, M. Gross, K. Meerholz, C. Bräuchle, K. Müllen, *Angew. Chem. Int. Ed.* **1998**, *37*, 1434. (b) U. Rohr, C. Kohl, K. Müllen, A. van de Draats, J. Warman, *J. Mater. Chem.* **2001**, *11*, 1789. (c) Z. An, J. Yu, B. Domercq, S. C. Jones, S. Barlow, B. Kippelen, S. R. Marder, *J. Mater. Chem.* **2009**, *19*, 6688. (d) H. A. Wegner, L. T. Scott, A. de Meijere, *J. Org. Chem.* **2003**, *68*, 883. (e) S. Müller, K. Müllen, *Chem. Commun.* **2005**, 4045. (f) Y. Avlasevich, C. Li, K. Müllen, *J. Mater. Chem.* **2010**, *20*, 3814.
17. (a) Y. Li, Z. Wang, *Org. Lett.* **2009**, *11*, 1385. (b) W. Jiang, H. Qian, Y. Li, Z. Wang, *J. Org. Chem.* **2008**, *73*, 7369.
18. Y. Li, J. Gao, S. D. Motta, F. Negri, Z. H. Wang, *J. Am. Chem. Soc.* **2010**, *132*, 4208.
19. (a) A. Tsuda, A. Osuka, *Science,* **2001**, *293*, 79. (b) S. Hiroto, A. Osuka, *J. Org. Chem.* **2005**, *70*, 4054. (c) A. Tsuda, H. Fruta, A. Osuka, *Angew. Chem. Int. Ed.* **2000**, *39*, 2549.
20. N. K. S. Davis, M. Pawlicki, H. L. Anderson, *Org. Lett.* **2008**, *10*, 3945.
21. C. Jiao, K.-W. Huang, J. Luo, K. Zhang, C. Chi, J. Wu, *Org. Lett.* **2009**, *11*, 4508.

22. W. Jiang, H. Qian, Y. Li, Z. Wang, *J. Org. Chem.* **2008**, *73*, 7369.
23. (a) A. L. Colmsjö, *Anal. Chem.* **1982**, *54*, 1673. (b) N. Nijegorodov, R. Mabbs, W. S. Downey, *Spectrochim. Acta, Part A* **2001**, *57*, 2673.
24. X.-T. Liu, J.-F. Guo, A.-M. Ren, S. Huang, J.-K. Feng, *J. Org. Chem.* **2012**, *77*, 585.
25. P. P. Ghoroghchian, P. R. Frail, K. Susumu, D. Blessington, A. K. Brannan, F. S. Bates, B. Chance, D. A. Hammer, M. J. Therien, *Proc. Natl. Acad. Sci. U.S.A.* **2005**, *102*, 2922.
26. A. Burke, L. Schmidt-Mende, S. Ito, M. Grätzel, *Chem. Commun.* **2007**, 234.
27. H. Suzuki, K. Ogura, N. Matsumoto, P. Prosposito, S. Schutzmann, *Mol. Cryst. Liq. Cryst.* **2006**, *444*, 51.
28. (a) L. Beverina, J. Fu, A. Leclercq, E. Zojer, P. Pacher, S. Barlow, E. W. Van Stryland, D. J. Hagan, J.-L. Brédas, S. R. Marder, *J. Am. Chem. Soc.* **2005**, *127*, 7282. (b) K. Kurotobi, K. S. Kim, S. B. Noh, D. Kim, A. Osuka, *Angew. Chem. Int. Ed.* **2006**, *45*, 3944.
29. J. Fabian, H. Nakazumi, M. Matsuoka, *Chem. Rev.* **1992**, *92*, 1197.
30. A. Mishra, R. K. Behera, P. K. Behera, B. K. Mishra, G. B. Behera, *Chem. Rev.* **2000**, *100*, 1973.
31. (a) J. Fabian, R. Zahradník, *Angew. Chem. Int. Ed.* **1989**, *28*, 677. (b) L. A. Ernst, R. K. Gupta, R. B. Mujumdar, A. S. Waggoner, *Cytometry*, **1989**, *10*, 3.
32. D. M. Sturmer, in *The Chemistry of Heterocyclic Compounds: Special Topics in Heterocyclic Chemistry*, A. Weissberger, E. C. Taylor (Ed.), John Wiley & Sons: New York, 1977.
33. A. E. Asato, D. T. Watanabe, R. S. H. Liu, *Org. Lett.* **2000**, *2*, 2559.
34. (a) K. Kundu, S.F. Knight, N. Willett, S. Lee, W. R. Taylor, N. Murthy, *Angew. Chem. Int. Ed.* **2009**, *48*, 299. (b) X. Chen, P. S. Conti, R. A. Moats, *Cancer Res.* **2004**, *64*, 8009. (c) Y. Lin, R. Weissleder, C. H. Tung, *Bioconjugate Chem.* **2002**, *13*, 605.
35. X. Chen, X. Peng, A. Cui, B. Wang, L. Wang, R. Zhang, *J. Photochem. Photobiol. A.* **2006**, *181*, 79.
36. A. A. Ischenko, N. A. Derevyanko, V. A. Svidro, *Opt. Spectrosc.* **1992**, *72*, 60.
37. N. Tyutyulkov, J. Fabian, A. Mehlhorn, F. Dietz, A. Tadjer, *Polymethine Dyes*, St. Kliment Ohridski University Press: Sofia, 1991.
38. K. Rurack, M. Kollmannsberger, J. Daub, *Angew. Chem. Int. Ed.* **2001**, *40*, 385.
39. K. Rurack, M. Kollmannsberger, J. Daub, *New J. Chem.* **2001**, *25*, 289.
40. H. Kim, A. Burghart, M. B. Welch, J. Reibenspies, K. Burgess, *Chem. Commun.* **1999**, 1889.
41. J. Chen, J. Reibenspies, A. Derecskei-Kovacs, K. Burgess, *Chem. Commun.* **1999**, 2501.
42. J. Chen, A. Burghart, A. Derecskei-Kovacs, K. Burgess, *J. Org. Chem.* **2000**, *65*, 2900.
43. A. Burghart, H. Kim, M. B. Welch, L. H. Thoresen, J. Reibenspies, K. Burgess, F. Bergstrom, L. B. A. Johansson, *J. Org. Chem.* **1999**, *64*, 7813.
44. G. Ulrich, R. Ziessel, A. Harriman, *Angew. Chem. Int. Ed.* **2008**, *47*, 1184.
45. R. Ziessel, G. Ulrich, A. Harriman, *New J. Chem.* **2007**, *31*, 496.
46. A. Loudet, K. Burgess, *Chem. Rev.* **2007**, *107*, 4891.
47. A. B. Descalzo, H. J. Xu, Z. Shen, K. Rurack, *Ann. N.Y. Acad. Sci.* **2008**, *1130*, 164.
48. Y.-H. Yu, Ana B. Descalzo, Z. Shen, M. Röhr, Q. Liu, Y.-W. Wang, M. Spieles, Y.-Z. Li, K. Rurack, X.-Z. You, *Chem. Asian. J.* **2006**, *1*, 176.
49. C. Murata, T. Masuda, Y. Kamochi, K. Todoroki, H. Yoshida, H. Nohta, M. Yamaguchi, A. Takadate, *Chem. Pharm. Bull.* **2005**, *53*, 750.

50. Z. Shen, H. Röhr, K. Rurack, H. Uno, M. Spieles, B. Schulz, G. Reck, N. Ono, *Chem. Eur. J.* **2004**, *10*, 4853.
51. S. Goeb, R. Ziessel, *Org. Lett.* **2007**, *9*, 737.
52. M. Wada, S. Ito, H. Uno, T. Murashima, N. Ono, T. Urano, Y. Urano, *Tetrahedron Lett.* **2001**, *42*, 6711.
53. Y. Yang, M. Lowry, X. Xu, J. O. Escobedo, M. Sibrian-Vazquez, L. Wong, C. M. Schowalter, T. J. Jensen, F. R. Fronczek, I. M. Warner, R. M. Strongin, *Proc. Natl. Acad. Sci.* **2008**, *105*, 8829.
54. Y. Yang, M. Lowry, C. M. Schowalter, S. O. Fakayode, J. O. Escobedo, X. Xu, H. Zhang, T. J. Jensen, F. R. Fronczek, I. M. Warner, R. M. Strongin, *J. Am. Chem. Soc.* **2006**, *128*, 14081.
55. Y. Yang, M. Lowry, C. M. Schowalter, S. O. Fakayode, J. O. Escobedo, X. Xu, H. Zhang, T. J. Jensen, F. R. Fronczek, I. M. Warner, R. M. Strongin, *J. Am. Chem. Soc.* **2007**, *129*, 1008.
56. G. Ulrich, S. Goeb, A. De Nicola, P. Retailleau, R. Ziessel, *Synlett.* **2007**, *2007*, 1517.
57. Y. Wu, D. H. Klaubert, H. C. Kang, Y.-Z. Zhang, U.S. Patent 6005113 (1999).
58. K. Umezawa, A. Matsui, Y. Nakamura, D. Citterio, K. Suzuki, *J. Am. Chem. Soc.* **2008**, *130*, 1550.
59. K. Umezawa, A. Matsui, Y. Nakamura, D. Citterio, K. Suzuki, *Chem. Eur. J.* **2009**, *15*, 1096.
60. (a) C. Jiao, K. Huang, J. Wu, *Org. Lett.* **2011**, *13*, 632. (b) C. Jiao, L. Zhu, J. Wu, *Chem. Eur. J.* **2011**, *17*, 6610.
61. L. Zeng, C. Jiao, X. Huang, K.-W. Huang, W.-S. Chin, J. Wu, *Org. Lett.* **2011**, *13*, 6026.
62. (a) M. A. T. Rogers, *J. Chem. Soc.* **1943**, 590. (b) W. H. Davis, M. A. T. Rogers, *J. Chem. Soc.* **1944**, 126. (c) E. B. Knott, *J. Chem. Soc.* **1947**, 1196.
63. (a) A. Loudet, K. Burgess, *Chem. Rev.* **2007**, *107*, 4891. (b) A. Gorman, J. Killoran, C. O'Shea, T. Kenna, W. M. Gallagher, D. F. O'Shea, *J. Am. Chem. Soc.* **2004**, *126*, 10619.
64. K. Flavin, K. Lawrence, J. Bartelmess, M. Tasior, C. Navio, C. Bittencourt, D. F. O'Shea, D. M. Guldi, S. Giordani, *ACS Nano* **2011**, *5*, 1198.
65. S. O. McDonnell, D. F. O'Shea, *Org. Lett.* **2006**, *8*, 3493.
66. W. Zhao, E. M. Carreira, *Angew. Chem. Int. Ed.* **2005**, *44*, 1677.
67. W. Zhao, E. M. Carreira, *Chem. Eur. J.* **2006**, *12*, 7254.
68. G. M. Fischer, M. K. Klein, E. Daltrozzo, A. Zumbusch, *Eur. J. Org. Chem.* **2011**, 3421.
69. G.M. Fischer, A. P. Ehlers, A. Zumbusch, E. Daltrozzo, *Angew. Chem. Int. Ed.* **2007**, *46*, 3750.
70. G. M. Fischer, M. Isomäki-Krondahl, I. Göttker-Schnetmann, E. Daltrozzo, A. Zumbusch, *Chem. Eur. J.* **2009**, *15*, 4857.
71. E. Daltrozzo, Uber den Einfluß Induktiver Mesomerer und Stenscher Effekte auf das Chemische und Spektroskopische Verhalten der Chinolyl Methane. PhD thesis, Technical University of Munich, 1965.
72. W. Sulger, Synthese and Spektroskopische Eigensceaften Neuartiger Fluorszenzfarbstoffe der Polymethinreihe. PhD thesis, University of Konstanz, 1981.
73. (a) H. Meier, U. Dullweber, *J. Org. Chem.* **1997**, *62*, 4821. (b) H. Meier, R. Petermann, J. Gerold, *Chem. Commun.* **1999**, 977. (c) H. Meier, R. Petermann, *Tetrahedron Lett.* **2000**, *41*, 5475. (d) J. Gerold, U. Holzenkamp, H. Meier, *Eur. J.*

Org. Chem. **2001**, *2001*, 2757. (e) R. Petermann, M. Tian, S. Tatsuura, M. Furuki, *Dyes Pigm.* **2003**, *57*, 43. (f) H. Meier, R. Petermann, *Helv. Chim. Acta.* **2004**, *87*, 1109. (g) J.-G. Chen, D.-Y. Huang, Y. Li, *Dyes Pigm.* **2000**, *46*, 93. (h) S.-H. Kim, J.-H. Kim, J.-Z. Cui, Y.-S. Gal, S.-H. Jin, K. Koh, *Dyes Pigm.* **2002**, *55*, 1. (i) S. Yagi, Y. Hyodo, S. Matsumoto, N. Takahashi, H. Kono, H. Nakazumi, *J. Chem. Soc. Perkin Trans. 1.* **2000**, 599. (j) Y. Hyodo, H. Nakazumi, S. Yagi, K. Nakai, *J. Chem. Soc. Perkin Trans. 1.* **2001**, 2823. (k) S. Yagi, S. Murayama, Y. Hyodo, Y. Fujie, M. Hirose, H. Nakazumi, *J. Chem. Soc. Perkin Trans. 1.* **2002**, 1417.

74. (a) A. Ajayaghosh, *Chem. Soc. Rev.* **2003**, *32*, 181. (b) A. Ajayaghosh, *Acc. Chem. Res.* **2005**, *38*, 449. (c) S. Sreejith, P. Carol, P. Chithra, A. Ajayaghosh, *J. Mater. Chem.* **2008**, *18*, 264.

75. Y. Chandrasekaran, G. K. Dutta, R. B. Kanth, S. Patil, *Dyes Pigm.* **2009**, *83*, 162.

76. (a) M. Kasha, *J. Chem. Phys.* **1952**, *20*, 71. (b) I. B. Berlmann, *J. Phys. Chem.* **1973**, *77*, 562. (c) M. Rae, A. Fedorov, M. N. Berberan-Santos, *J. Chem. Phys.* **2003**, *119*, 2223.

77. U. Mayerhöffer, B. Fimmel, F. Würthner, *Angew. Chem. Int. Ed.* **2012**, *51*, 164.

78. M. Tian, S. Tatsuura, M. Furuki, Y. Sato, I. Iwasa, L. S. Pu, *J. Am. Chem. Soc.* **2003**, *125*, 348.

79. H. Langhals, *Angew. Chem. Int. Ed.* **2003**, *42*, 4286.

80. K. Yesudas, K. Bhanuprakash, *J. Phys. Chem. A*, **2007**, *111*, 1943.

81. A. L. Puyad, Ch. Prabhakar, K. Yesudas, K. Bhanuprakash, V. J. Rao, *J. Mol. Struc. Theochem*, **2009**, *904*, 1.

82. M. P. Le Paillard, A. Robert, C. Garrigou-Lagrange, P. Delhaes, P. Le Maguerès, L. Ouahab, L. Toupet, *Synth. Met.* **1993**, *58*, 223.

83. P. R. Ashton, V. Balzani, J. Becher, A. Credi, M. C. T. Fyfe, G. Mattersteig, S. Menzer, M. B. Nielsen, F. M. Raymo, J. F. Stoddart, M. Venturi, D. J. Williams, *J. Am. Chem. Soc.* **1999**, *121*, 3951.

84. R. M. Metzger, *J. Mater. Chem.* **1999**, *9*, 2027.

85. M. Bendikov, F. Wudl, D. F. Perepichka, *Chem. Rev.* **2004**, *104*, 4891.

86. D. F. Perepichka, M. R. Bryce, C. Pearson, M. C. Petty, E. J. L. McInnes, J. P. Zhao, *Angew. Chem. Int. Ed.* **2003**, *42*, 4636.

87. C. Jia, S.-X. Liu, C. Tanner, C. Leiggener, A. Neels, L. Sanguinet, E. Levillain, S. Leutwyler, A. Hauser, S. Decurtins, *Chem. Eur. J.* **2007**, *13*, 3804.

88. S.-i. Kato, T. Matsumoto, T. Ishi-i, T. Thiemann, M. Shigeiwa, H. Gorohmaru, S. Maeda, Y. Yamashita, S. Mataka, *Chem. Commun.* **2004**, 2342.

89. M. Velusamy, K. R. J. Thomas, J. T. Lin, Y. S. Wen, *Tetrahedron Lett.* **2005**, *46*, 7647.

90. G. Qian, B. Dai, M. Luo, D. Yu, J. Zhan, Z. Zhang, D. Ma, Z. Y. Wang, *Chem. Mater.* **2008**, *20*, 6208.

91. M. Karikomi, C. Kitamura, S. Tanaka, Y. Yamashita, *J. Am. Chem. Soc.* **1995**, *117*, 6791.

92. K. Ono, S. Tanaka, Y. Yamashita, *Angew. Chem. Int. Ed. Engl.* **1994**, *33*, 1977.

93. R. Lakowicz, *Principles of Fluorescence Spectroscopy*, Kluwer: New York, 1999.

94. G. Qian, Z. Zhong, M. Luo, D. Yu, Z. Zhang, D. Ma, Z. Y. Wang, *J. Phys. Chem. C.* **2009**, *113*, 1589.

95. Y. Yang, R. T. Farley, T. T. Steckler, S.-H. Eom, J. R. Reynolds, K. S. Schanze, J. Xue, *Appl. Phys. Lett.* **2008**, *93*, 163305.

96. T. T. Steckler, K. A. Abboud, M. Craps, A. G. Rinzler, J. R. Reynolds, *Chem. Commun.* **2007**, 4904.

97. G. Qian, Z. Zhong, M. Luo, D. Yu, Z. Zhang, Z. Y. Wang, D. Ma, *Adv. Mater.* **2009**, *21*, 111.

98. G. Qian, Z. Y. Wang, *Can. J. Chem.* **2010**, *88*, 192.

99. G. C. Welch, R. Coffin, J. Peet, G. C. Bazan, *J. Am. Chem. Soc.* **2009**, *131*, 10802.

100. J. C. Beeson, L. J. Fitzgerald, J. C. Gallucci, R. E. Gerkin, J. T. Rademacher, A. W. Czarnik, *J. Am. Chem. Soc.* **1994**, *116*, 4621.

101. P. Secondo, F. Fages, *Org. Lett.* **2006**, *8*, 1311.

102. T. Ishi-i, T. Hirayama, K.-i. Murakami, H. Tashiro, T. Thiemann, K. Kubo, A. Mori, S. Yamasaki, T. Akao, A. Tsuboyama, T. Mukaide, K. Ueno, S. Mataka, *Langmuir.* **2005**, *21*, 1261.

103. T.-H. Chang, B.-R. Wu, M. Y. Chiang, S.-C. Liao, C. W. Ong, H.-F. Hsu, S.-Y. Lin, *Org. Lett.* **2005**, *7*, 4075.

104. T. Ishi-i, K.-i. Murakami, Y. Imai, S. Mataka, *Org. Lett.* **2005**, *7*, 3175.

105. J. Hu, D. Zhang, S. Jin, S. Z. D. Cheng, F. W. Harris, *Chem. Mater.* **2004**, *16*, 4912.

106. B. R. Kaafarani, L. A. Lucas, B. Wex, G. E. Jabbour, *Tetrahedron Lett.* **2007**, *48*, 5995.

107. M. Luo, H. Shadnia, G. Qian, X. Du, D. Yu, D. Ma, J. S. Wright, Z. Y. Wang, *Chem. Eur. J.* **2009**, *15*, 8902.

108. B. Gao, Synthesis and Characterization of Electron-Deficient Pyrazine Conjugated Oligomers, Ph.D. Dissertation, Institute of Applied Chemistry, Chinese Academy of Sciences, P. R. China, 2006.

109. S. Kohlmann, S. Ernst, W. Kaim, *Angew. Chem. Int. Ed. Engl.* **1985**, *24*, 684.

110. T. Shida, *Electronic Absorption Spectra of Radical Ions*, Physical Sciencs Data 34, Elsevier: Amsterdam, 1988.

111. M. Haga, E. S. Dodsworth, A. B. P. Lever, *Inorg. Chem.* **1986**, *25*, 447.

112. (a) L. F. Joulié, E. Schatz, M. D. Ward, F. Weber, L. J. Yellowlees; *J. Chem. Soc., Dalton Trans.* **1994**, 799. (b) A. M. Barthram, R. L. Cleary, R. Kowallick, M. D. Ward, *Chem. Commun.* **1998**, 2695. (c) D. A. Shukla, B. Ganguly, P. C. Dave, A. Samanta; A. Das, *Chem. Commun.* **2002**, 2648.

113. P. F. H. Schwab, S. Diegoli, M. Biancardo, C. A. Bignozzi, *Inorg. Chem.* **2003**, *42*, 6613.

114. J. Garcia-Canadas, A. P. Meacham, L. M. Peter, M. D. Ward, *Angew. Chem., Int. Ed.* **2003**, *42*, 3011.

115. (a) M. D. Ward, J. A. McCleverty, *J. Chem. Soc., Dalton Trans.* **2002**, 275. (b) C. G. Pierpont, *Coord. Chem. Rev.* **2001**, *216*, 99. (c) A. B. P. Lever, H. Masui, R. A. Metcalfe, D. J. Stufkens, E. S. Dodsworth, P. R. Auburn, *Coord. Chem. Rev.* **1993**, *125*, 317.

116. (a) D. M. Brown, *Mixed Valence Compounds*, D. Reidel: Dordrecht, Holland, 1980. (b) K. Prassides, *Mixed Valence Systems: Applications in Chemistry, Physics and Biology*, NATO ASI Series 343; Kluwer Academic Publishers: Dordrecht, The Netherlands, 1990.

117. J. Woodward, *Philos. Trans. R. Soc. London* **1724**, 33, 15.

118. M. B. Robin, P. Day, *Adv. Inorg. Chem. Radiochem.* **1967**, *10*, 247.

119. G. C. Allen, N. S. Hush, *Prog. Inorg. Chem.* **1967**, *8*, 357.

120. (a) C. Creutz, H. Taube, *J. Am. Chem. Soc.* **1969**, *91*, 3988. (b) C. Creutz, H. Taube, *J. Am. Chem. Soc.* **1973**, *95*, 1086.

121. R. J. Crutchley, *Adv. Inorg. Chem.* **1995**, *41*, 273.

122. (a) K. D. Demadis, D. C. Hartshorn, T. J. Meyer, *Chem. Rev.* **2001**, *101*, 2655. (b) R. Rocha, F. N. Rein, H. Jude, A. P. Shreve, J. J. Concepcion, T. J. Meyer, *Angew. Chem. Int. Ed.* **2008**, *47*, 503. (c) J. J. Concepcion, D. M. Dattelbaum, T. J. Meyer, C. Rocha, *Phil. Trans. R. Soc. A.* **2008**, *366*, 163.

123. (a) W. Kaim, A. Klein, M. Glöckle, *Acc. Chem. Res.* **2000**, *33*, 755. (b) W. Kaim, B. Sarkar, *Coord. Chem. Rev.* **2007**, *251*, 584. (c) W. Kaim, *Coord. Chem. Rev.* **2011**, *255*, 2503.

124. T. Scheiring, W. Kaim, J. A. Olabe, A. R. Parise, J. Fiedler, *Inorg. Chim. Acta,* **2000**, *300–302*, 125.

125. (a) F. Felix, U. Hauser, H. Siegenthaler, F. Wenk, A. Ludi, *Inorg. Chim. Acta,* **1975**, *15*, L7. (b) M. Ketterle, W. Kaim, J. A. Olabe, A. R. Parise, J. Fiedler, *Inorg. Chim. Acta,* **1999**, *291*, 66.

126. (a) W. Bruns, W. Kaim, E. Waldhör, M. Krejcik, *J. Chem. Soc. Chem. Commun.* **1993**, 1868. (b) W. Bruns, W. Kaim, *J. Organomet. Chem.* **1990**, *390*, C45.

127. (a) W. Kaim, V. Kasack, H. Binder, E. Roth, J. Jordanov, *Angew. Chem. Int. Ed. Engl.* **1988**, *27*, 1174. (b) W. Kaim, V. Kasack, *Inorg. Chem.* **1990**, *29*, 4696. (c) V. Kasack, W. Kaim, H. Binder, J. Jordanov, E. Roth, *Inorg. Chem.* **1995**, *34*, 1924.

128. Y. Qi, P. Desjardins, Z. Y. Wang, *J. Opt. A: Pure Appl. Opt.* **2002**, *4*, 1.

129. Y. Qi, Z. Y. Wang, *Macromolecules,* **2003**, *36*, 3146.

130. M. F. Rastegar, E. K. Todd, H. Tang, Z. Y. Wang, *Org. Lett.* **2004**, *6*, 4519.

131. S. H. Kim, M. Matsuoka, M. Yomoto, Y. Tsuchiya, T. Kitao, *Dyes Pigm.* **1987**, *8*, 388.

132. U. Lehmann, D. Heizler, WO 2008/086931 A1 (July 24, 2008).

133. K. Arumugam, M. C. Shaw, P. Chandrasekaran, D. Villagrán, T. G. Gray, J. T. Mague, J. P. Donahue, *Inorg. Chem.* **2009**, *48*, 10591.

134. (a) F. Holtrup, G. Müller, H. Quante, S. de Feyter, F.C. De Schryver, K. Müllen, *Chem. Eur. J.* **1997**, *3*, 219. (b) H. Quante, K. Müllen, *Angew. Chem. Int. Ed. Engl.* **1995**, *34*, 1323.

135. G. M. Fischer, E. Daltrozzo, A. Zumbusch, *Angew. Chem. Int. Ed.* **2011**, *50*, 1406.

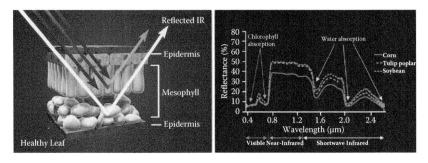

Color Figure 1.6 Reflection of a green leaf, chlorophyll, and other plants. (From http://missionscience.nasa.gov/ems/08_nearinfraredwaves.html.)

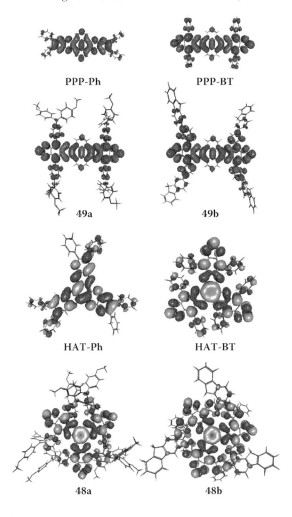

Color Figure 3.9 Comparison of LUMO spatial distribution (isovalue = 0.01 a.u) between reference chromophores PPP-Ph, PPP-BT, HAT-Ph, and HAT-BT and synthesized chromophores **48** and **49**.

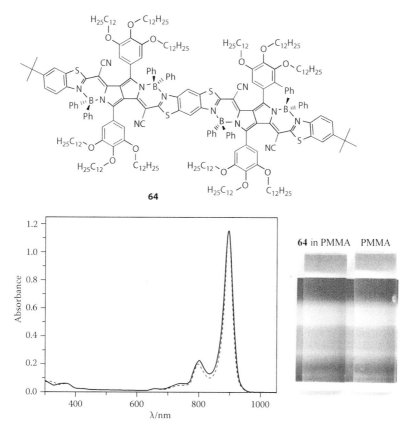

Color Figure 3.13 DCH-Ru trimers and their solutions in the Ru(II)/Ru(II) and Ru(II)/Ru(III) states.

64

Color Figure 3.17 Absorption spectra of **64** in chloroform (dotted line) and in PMMA (solid line). A photograph of the PMMA-cuboid containing **64** (left) and the PMMA-cuboid without dye (right). (Adapted with permission from G. M. Fischer et al., *Angew. Chem. Int. Ed.* **2011**, *50*, 1406.)

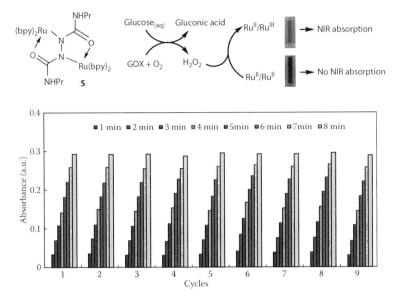

Color Figure 5.12 Top: Proposed detection of glucose or hydrogen peroxide using dinuclear ruthenium complex. Cuvettes shown are ACN solutions of complex **5** in its respective oxidation states. Bottom: Normalized absorbance at 1150 nm of a film containing **5** on ITO at a given time after treatment with H_2O_2 (0.3 M) in TRIS solution. Film was reduced electrochemically (–0.5 V) in 9 cycles. (Adapted with permission from S. Xun et al., *Org. Lett.* **2006**, *8*, 1697. Copyright (2006) American Chemical Society.)

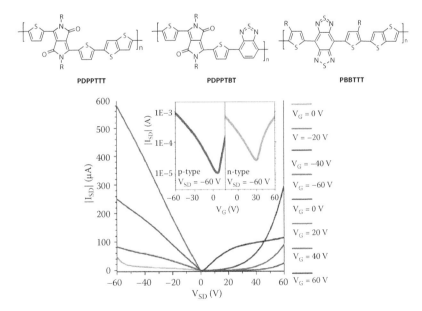

Color Figure 5.23 p-Type **PDDPTTT** and ambipolar **PDDPTBT** and **PBBTTT** polymers. Transfer (inset) and output characteristics of a **PBBTTT** transistor. (Adapted with permission from J. Fan et al., *Adv. Mater.* **2012**, *24*, 2186.)

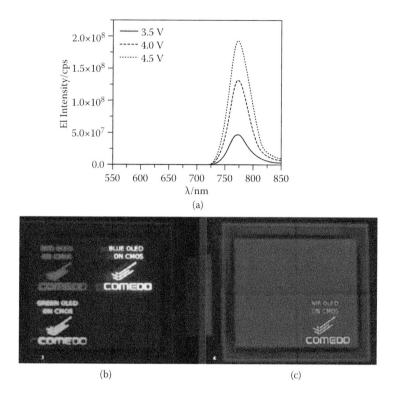

Color Figure 5.27 (a) EL spectra of an NIR OLED at different applied bias. (b) Detailed image of an RGB-NIR microdisplay recorded in the visible range of light. (c) Detailed image of RGB-NIR microdisplay recorded in the NIR region of light.

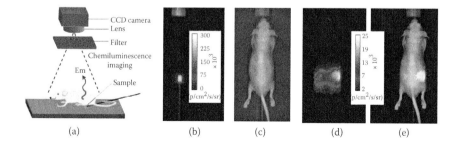

Color Figure 5.28 Chemiluminescence from squaraine rotaxane endoperoxide (SREP) at 38°C penetrates through a living nude mouse. (a) Experimental setup for planar chemiluminescence. (b) Chemiluminescence pixel intensities from a small tube containing SREP (250 nmol) in $C_2D_2Cl_4$. (c) Photographs of mouse located above the tube. (d,e) Pixel intensity map of chemiluminescence that is transmitted through the mouse (target background ratio = 11.6). (Adapted with permission from J. M. Baumes et al., *Nature Chem.* **2010**, *2*, 1025.)

chapter four

Near-infrared absorbing polymers

In the law of God, there is no statute of limitations.

Robert Louis Stevenson (1850–1894)

Near-infrared (NIR) absorbing polymers can be classified as (1) nonconjugated polymers having the isolated NIR chromophoric units in the main chain or side chain and (2) conjugated polymers containing the segments with an effective conjugation length responsible for the NIR absorption. For the first class of NIR polymers, NIR chromophores need to be functionalized with the reactive groups suitable for polymerization or grafting usually via the formation of C-C and C-O bonds. For making conjugated NIR-absorbing polymers, the AB or AA and BB type of monomers are selected for polycondensation usually by the metal-catalyzed homo- or cross-coupling reactions, such as the Suzuki and Stille cross-coupling reactions, or by electrochemical polymerization. The cross-coupling reactions can be applied to a variety of monomers bearing many different functional groups and are able to produce high molecular weight pure polymers. Electrochemical polymerization is suitable for polymerization of redox-active monomers that usually bear the thiophene or pyrrole group and making a final polymer film on the electrode. However, due to residual doping, the band gap of the polymer is difficult to be accurately determined and is often reported to be unrealistically low.

In a conjugated polymer system, the HOMO and LUMO levels of the repeat units (monomers) disperse into valence and conduction bands upon chain extension. Dispersion magnitudes, or the bandwidth of the monomer HOMO and LUMO levels, are represented by W_L and W_H (Figure 4.1). The magnitudes of W_L and W_H are strongly dependent on the degree of overlap between the atomic orbitals on the coupling positions of the consecutive units. The maximal values for W_L and W_H are only reached in the case of an unobstructed overlap. Therefore, the monomers must be well designed, so that they can lead to the formation of a low band-gap repeat unit or even longer segment with effective overlapping. Thus, the energy gap (E_g) of the conjugated polymer is usually smaller than that of monomers and, accordingly, the absorption of polymers is redshifted in comparison with monomers. However, deviation from this ideal situation can occur when (1) steric hindrance forces the consecutive aryl or

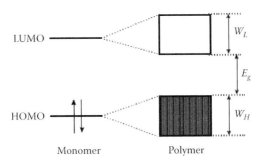

Figure 4.1 Dispersion magnitudes W_H and W_L of the monomer HOMO and LUMO levels, respectively, upon chain extension.

heteroaryl units out of plane, or (2) the size of the atomic orbitals at the coupling positions is diminished. There are many conjugated polymers that are deemed to be NIR absorbing but actually not. For example, some conjugated donor–acceptor polymers containing the electron-donating group of pyrrole and/or thiophene and the electron-accepting group of a cyanosubstituted aryl unit were prepared by electrochemical oxidation (acetonitrile/NBu₄ClO₄) of their corresponding monomers [1]. The band gaps of these polymers were estimated to be in a range of 2.0 and 2.7 eV. The large band gap may well be due to the diminished size of the atomic orbitals at the coupling sites between the donor and acceptor units.

1.6 eV   **2.0 eV**   **2.2 eV**

For nonconjugated NIR polymers, the band gap levels or absorption of the monomers and the resulting polymers are basically the same. Polymers offer a processing advantage over small molecules while keeping the same or even better optical properties for certain applications, especially for thin-film device applications. Therefore, the main reason for grafting NIR dyes onto a nonconjugated polymer is for thin-film application by spin coating or casting. However, by doping NIR dyes in a host polymer, one can easily and inexpensively obtain a wide range of NIR dye-doped polymers for almost the same applications as nonconjugated NIR polymers. Ultimately, NIR polymers should combine the physical properties of polymers (low specific weight, processibility, solubility, tunable mechanical properties, flexibility, etc.) with those of semiconductors to obtain unique and novel materials with numerous exciting applications.

4.1 Nonconjugated NIR-absorbing polymers

In principle, any NIR chromophore can be functionalized as a mono-mer to be incorporated into a polymer in one way or another. A common method is to functionalize a chromophore with an acrylate or vinyl group for addition polymerization. In most cases, NIR chromophore molecules need to be chemically modified to have a vinyl, acrylate, or other poly-merizable group as monomers. Free-radical polymerization is typically used to produce (co)polyarylates, (co)polyacrylamides, or vinyl poly-mers containing the desired chromophores as pendent groups. A more practical approach to nonconjugated NIR polymers is based on polymer grafting. A polymer containing a reactive pendent group (e.g., -OH, -NH$_2$, halo, and epoxide) is typically used for grafting with NIR chromophores bearing a suitable reactive group. Similarly, a precursor route to the target NIR polymers is sometimes used, especially in the case that the required NIR chromophore monomers are difficult to make. A precursor polymer is prepared first and then the NIR chromophore is selectively introduced by a specific reaction.

DCH-Ru

R,R′ = alkyl,
aryl, alkoxy, etc.

PDCH-Ru

The precursor strategy was successfully applied to the synthesis of non-conjugated polymers containing pendent NIR metal complex chromo-phores. **DCH-Ru** complexes are known to absorb strongly between 1000 to 1800 nm when being in the mixed-valence state and have PL emission at 790 nm due to triplet metal–ligand charge transfer (^{3}MLCT) [2]. To further explore the NIR electrochromic and electroluminescent proper-ties in the solid-state devices, DCH-Ru polymers are desirable since the polymers can offer some advantages over DCH-Ru small molecules, such as better processing capability, film quality, and better control over mor-phology. Usually, the target polymers are synthesized by polymerization of a DCH-Ru acrylate or vinyl monomer with or without a comonomer.

DCH-Ru complexes are readily synthesized in good yields from dicarbonylhydrazines and Ru(bpy)$_2$Cl$_2$ in alkaline water [3]. However, it is not so straightforward to synthesize a vinyl or acrylic-containing DCH-Ru monomer, because the vinyl or acrylic group in the dicarbonylhydrazine ligand is quite labile to a base and does not survive in the strong alkaline medium during the complexation reaction with Ru(bpy)$_2$Cl$_2$. Therefore, a precursor route was used to synthesize a series of NIR electrochromic and electroluminescent polymers (**PDCH-Ru**) containing the pendant DCH-Ru chromophore [2d]. The ligand copolymers were first prepared, which were subsequently treated with Ru(bpy)$_2$Cl$_2$. Unlike **DCH-Ru** complexes, these polymers show good thermal stability and film-forming ability; Similar to **DCH-Ru** complexes, **PDCH-Ru** polymers are NIR electrochromic, displaying an intense absorption centered at 1600 nm upon oxidation to the mixed-valence state. Single-layer diode devices comprising a layer of the complex polymers sandwiched between the ITO and Au electrodes emitted the NIR light centered at 790 nm at ambient temperature.

Nickel bis(dithiolenes) are a class of very strong NIR chromophores that possess an extremely intense NIR transition as low as 7000 cm^{-1} (~1428 nm) with molar extinction coefficients as high as 80,000 M^{-1}cm^{-1}. For example, the absorption spectrum of [Ni(S$_2$C$_2$Me$_2$)$_2$]^0 in acetonitrile displays a single intense low-energy transition with a maximum near 800 nm; the reduced mono-anionic [Ni(S$_2$C$_2$Me$_2$)$_2$]$^{1-}$ also absorbs the NIR light at the wavelength slightly above 800 nm, and a similar low-energy absorption feature in the reduced dianionic [Ni(S$_2$C$_2$Me$_2$)$_2$]$^{2-}$ complex is absent completely. Nickel bis(dithiolenes) have a square-planar coordination structure, which is quite rigid and "ribbonlike." As a consequence, these NIR metal complexes and their fully conjugated polymers, such as poly(metal tetrathiooxalates) [4], poly(metal ethylenetetrathiolates) [5], poly(metal tetrathiosquarates) [6], poly(metal tetrathiafulvalenetetrathiolates) [7], poly(metal tetrathianaphthalenes) [8], and poly(metal benzenetetrathiolates) [9], are insoluble in common organic solvents and are difficult to form thin films for device applications.

Ni(S$_2$C$_2$Me$_2$)$_2$

λ_{max} = 800 nm

P[Ar(S$_2$C$_2$)$_2$]
A = O, S, CH$_2$, (CH$_2$)$_{10}$, (CH$_2$)$_{22}$, or (OCH$_2$CH$_2$)$_3$O
λ_{max} = 900–940 nm

In order to circumvent the low solubility of these fully conjugated polymers, Wang and Reynolds incorporated square planar nickel bis(dithiolenes) into the main chain of nonconjugated polymers that contain flexible units [10]. A variety of flexible linkages in **P[Ar(S$_2$C$_2$)$_2$]** were utilized to separate the nickel bis(dithiolene) complexes and include -O-, -S-, -CH$_2$-, -(CH$_2$)$_{10}$-, -(CH$_2$)$_{22}$-, and -(OCH$_2$CH$_2$)$_3$O-. The polymers with short flexible linkages in this series are highly soluble in both aqueous and organic solvents in the reduced (dianionic) form and are slightly soluble in the oxidized or neutral form. Increasing the length of the organic flexible linkage in the polymer main chain increases the solubility of **P[Ar(S$_2$C$_2$)$_2$]** in the oxidized form. Three different oxidation states of the nickel complex were observed electrochemically with the [NiL$_2$$^{2-}$]$_n$ ⇆ [NiL$_2$$^{1-}$]$_n$ and [NiL$_2$$^{1-}$]$_n$ ⇆ [NiL$_2$^{0}]$_n$ redox processes. All of the polymer solutions show a absorption maximum at ca. 900 nm, which is assigned to the 2b$_{1u}$ ↔ 3b$_{2g}$ transition [11]. The ability to directly cast films of polymers containing these NIR absorbers may ultimately lead to films significantly more homogeneous than dye-doped polymers [12].

NIR chromophores can also be grafted onto a polymer via a selective reaction. NIR-cyanine dye was used as a fluorescent emitter and introduced into a well-defined, water-soluble dendrimer for biomaterials applications [13]. The dendrimer is multifunctionalized in a highly controllable and orthogonal fashion. By the reaction of the amine termini with a chloro-cyanine dye, the NIR dyes are grafted at the dendrimer's periphery and the dye-functionalized dendrimer shows fluorescence in the NIR region with a large Stokes shift and relatively high quantum yields and no toxicity toward T98G human cells.

4.2   Conjugated quinoid type polymers

Minimizing the bond length alternation along the backbone of a conjugated polymer is an important guideline in band-gap reduction and realization of NIR-absorbing polymers. There are two ways to increase the double-bond character between the repeating units of a conjugated polymer in order to reduce bond-length alternation; one is to go from the aromatic to the quinoid state, and another is to introduce a strong electron donor (D) and a strong electron acceptor (A). In the first approach, band-gap reductions in a conjugated quinoid type polymer are accomplished by making the quinoidal structure energetically more favorable due to a Peierls effect [14], as is the case in polyisothianaphthene [15]. In the second route, the interaction between a strong donor and a strong acceptor can also give rise to an increased double-bond character between these units, since they can accommodate the charges that are associated with such a mesomerism (D–A ↔ D$^+$ = A$^-$). Hence, a conjugated D–A type polymer with an alternating sequence of the appropriate donors and acceptors in the main chain is likely to have a decreased band gap.

Polythiophene is a typical conjugated polymer and has a high band gap of ~2 eV. By fusing benzene onto the thiophene, polyisothianaphthene (PITN) has a much lower band gap of roughly 1 eV (λ_{max} = 775 nm, film) [16]. The main reasons for this large difference in band gap is due to the difference in contribution of the quinoid structure to the ground state of the polymer. In polythiophene, such a contribution is rather small and energetically unfavorable, resulting in a large single bond character of the thiophene–thiophene linkage and hence a large bond-length alternation. Driven by the gain in aromaticity of the fused benzene ring, the quinoidal structure in PITN is more energetically favorable, resulting in a large double-bond character of the thiophene–thiophene linkage [15].

Many other polymers representing structural variations on the isothianaphthene unit have been reported, such as poly(dialkoxy isothianaphthene)s [17], poly(dialkyl thienopyrazine)s [18], and various fused thienothiophene polymers [19]. For example, a solution processable poly(2,3-dihexylthieno[3,4-b]pyrazine) had a band gap of 0.95 eV (calculated from electrochemical data) and showed a very broad absorption from 300 nm to 1250 nm with a maximum at 915 nm as film [20]. A film of poly(2-decylthieno[3,4-b]thiophene-4,6-diyl) showed a NIR absorption with λ_{max} of 925 nm and an optical band gap of 0.92 eV [21]. The same strategy was applied to a conjugated hydrocarbon polymer, poly(**indenofluorene**), composed of the 3,9-di-*tert*-butylindeno[1,2-b]fluorine unit [22]. The molecular design is based on a hypothesis that the aromaticity of the central benzene ring is lost on going from the structure with a large double-bond character to the structure with a large single-bond character between the units. In this case, it is highly possible that the presence of a significant torsion about exocyclic double bonds (left structure) makes the quinoid state (right structure) a preferred structure that contributes to the electronic ground state, thus reducing the band-gap energy. In fact, this polymer displays the lowest energy gap (1.55 eV) known for a well-defined neutral hydrocarbon polymer. The longest wavelength absorption is shifted into the NIR up to 799 nm.

poly(indenofluorene)

4.3 Conjugated dye-containing polymers

One practical approach to NIR-absorbing conjugated polymers is based on the use of known organic dyes with inherently low HOMO–LUMO energy gap as building blocks. As discussed in Chapter 3, a variety of NIR-absorbing organic dyes are known, such as squaraines, croconaines, porphyrins, rylenes, cyanines, diketopyrrolopyrroles (DPP), and BODIPY, but only a few have been successfully utilized as a building block to construct NIR-absorbing conjugated polymers.

4.3.1 Polysquaraines and polycroconaines

Squaraines are one of such NIR dyes that are suitable for the design of polysquaraines with low band gaps, by virtue of favorable optical properties, immense flexibility for synthetic manipulation, and theoretical calculations [23]. Calculations have shown that the hybridization of the energy levels of the donor and the acceptor, particularly the high-lying HOMO of the donor fragment and the low-lying LUMO of the acceptor fragment, yields a D–A monomer with an unusually low HOMO–LUMO separation. Further hybridization upon chain extension then converges to the low band gaps. For example, the calculated band gap of polymer **1** is 0.5 eV, whereas the calculated band gap of polymer **2** is even as low as 0.2 eV.

The early attempts toward this class of low band-gap polymers made by polycondensation of squaric acid with N-alkylcarbazoles were not promising, because the insoluble and intractable nature of the resulted black powder materials were hardly characterized and determined for their band-gap values [24]. Later, benzobisthiazoles and benzobispyrrolines

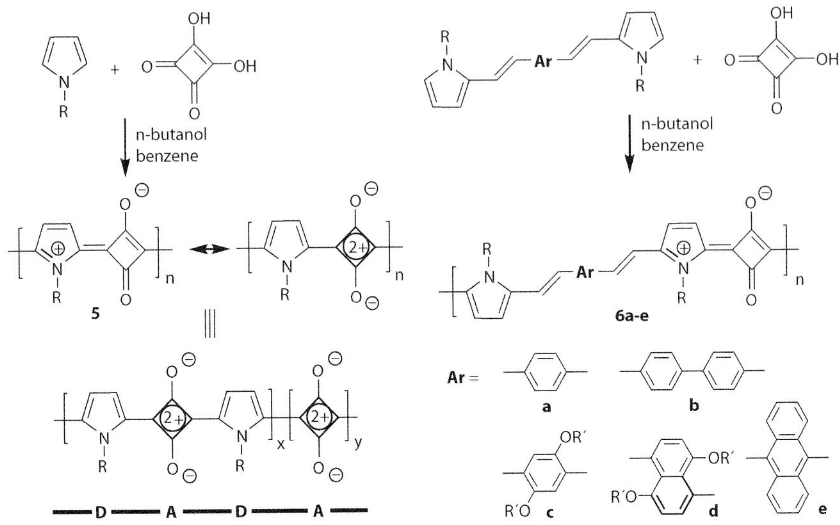

Scheme 4.1 Polycondensation of squaric acid with pyrrole and bispyrrole monomers.

were shown to condense with squaric acid to form water-soluble polysquaraines **3** and **4** with low optical band gaps of 0.7 and 1.2 eV, respectively [25]. Lynch et al. [26] and Ajayaghosh et al. [27] independently explored the pyrrole derivatives as monomers in the synthesis of polysquaraines and opened up a synthetic avenue to a large number of polysquaraines that show intense NIR absorption, NIR fluorescence, good semiconductivity, cation sensing, and other unique properties [28]. The simplest polysquaraines, such as polymer **5,** is readily derived from the reaction of pyrrole and squaric acid (Scheme 4.1). However, this polymer fails to meet the expectation of significant bathochromic shift in absorption, as its absorption maximum of 581 nm (Figure 4.2) is not much different from the corresponding monomeric dye units. Thus, a mere extension of conjugation in the pyrrole–squaric acid system is not sufficient to lower the band gaps. Considering the unit derived from squaric acid as an electron acceptor and the pyrrole unit as an electron donor, the structure of polysquaraines **5** can be viewed to consist of a D–A–D repeating unit, in which A is an electron-deficient cyclobutene moiety. Therefore, a large band gap of polymer **5** is a result of weak D–A–D interactions of the squaraine unit brought by the highly electron-deficient bridging unit or a mismatch of donor and acceptor strength due to relatively much weaker donor strength of pyrrole. Based on this rationale, it is reasonable to argue that introduction of an electron-rich bridging group should strengthen the D–A–D interaction or enhance the donor strength of electron-rich moiety, which leads to a quinoid-type resonance structure, thereby improving the effective

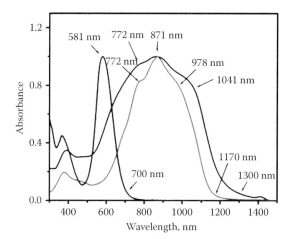

Figure 4.2 Absorption spectra of polysquaraines **5** (λ_{max} = 581 nm) and **6c** when R = CH$_3$ (broad absorption tailing at 1300 nm) and R = n-C$_{12}$H$_{25}$ (tailing at 1170 nm). (Adapted with permission from A. Ajayaghosh, *Acc. Chem. Res.* **2005**, *38*, 449.)

conjugation length and thus lowering the optical band gap. Accordingly, a series of NIR-absorbing polysquaraines **6a-e** were prepared by polycondensation of electron-rich bispyrroles with squaric acid (Scheme 4.1) [29]. Polymers **6** have a band-gap level around 1.0 eV and absorb broadly with a maximum around 850–1040 nm.

It was found that the absorption spectra of polymers **6** are strongly influenced by the electronic properties of the bridging unit and the length of the aliphatic side chains of the polymer backbone. For example, polysquaraines **6c** showed a broad absorption with several shoulder peaks when the side chain on the pyrrole nitrogen atom is short (R = CH$_3$), but a relatively narrow band when R is n-C$_{12}$H$_{25}$ (Figure 4.2). The broad window of the absorption spectra starting from 500 nm down to 1200 nm in these cases is quite remarkable when compared to the narrow absorption spectrum with a maximum at 581 nm of polysquaraines **5** (Figure 4.2). The strong D–A interaction in polymers **6** is likely to facilitate the resonance-stabilized quinoid structures, which is responsible for the observed NIR absorption.

The chemistry of croconic acid or 4,5-dihydroxycyclopentenetrione is similar to squaric acid in terms of the reactivity toward nucleophiles. As a higher homologue of squaric acid, croconic acid can also react with benzobisthiazoles and benzobispyrrolines to form the corresponding polycroconaines **7a-c**. Polymer **7b** shows a band gap of 0.45 eV (absorption edge), which is relatively lower than that of analog **7a** [25a,b]. Due to the better D–A interactions, polycroconaines show lower band gaps in comparison with polysquaraines. A comparison of the solid-state absorption spectra of water-soluble polysquaraines **3** and **4** and polycroconaine **7c** with photoconductivity properties indicate that the absorption gives rise to charge carriers and hence the absorption edge coincides with the band gap between the conduction and valence bands [25c].

4.3.2 *Conjugated polymers based on other organic dyes*

Depending on their structures, DPP, BODIPY, porphyrin, and rylene dyes can absorb the light in the NIR region and are synthetically feasible and available for use as building blocks to construct higher oligomers and polymers. By using appropriate π-conjugated connecting units, low band-gap conjugated polymers can readily be prepared and show intense NIR absorption and other unique optical and electrical properties.

	λ_{max}(nm)
8a	656 (CHCl$_3$)
8b	660 (CHCl$_3$)
8c	700 (film)
8d	780 (CHCl$_3$)
8e	890 (film)

Diketopyrrolo[3,4-*c*]pyrrole (DPP) is a key structural unit in an important class of red pigments that have exceptional light, weather, and heat stability [30], and is also widely used as a building block in construction of a variety of compounds and polymers for device applications, in particular light-emitting diodes [31], photovoltaic solar cells [32], and thin-film transistors [33]. DPP is a red chromophore and absorbs the light below 650 nm. By incorporating DPP into a D–A conjugated system, a bathochromic shift can be realized and the band gap of DPP-based polymers depends on the structures of donor monomers used in polymerization. By changing the donor strength and π-conjugation, the absorption of DPP-based polymers can be tuned. Polymers **8a-c** showed absorption bands in a range of 500–700 nm, while **8d,e** exhibited broad absorption in a range of 500–1000 nm with a maximum in the NIR spectral region, clearly showing the effect of the energy levels of donor segment on the band gaps of polymers [34].

9a, λ_{max} = 812 nm (CHCl$_3$) **9b**, λ_{max} = 857 nm (CHCl$_3$) **9c**, λ_{max} = 780 nm (film)

The fused thieno[3,2-*b*]thiophene (TT) unit has been used extensively in many low band-gap conjugated polymers that typically show high charge carrier mobility [35]. The TT units can extend the polymer coplanarity and also promote a more delocalized HOMO distribution along the polymer backbone, which is expected to enhance intermolecular charge carrier hopping. One approach to the inclusion of TT units in DPP-based polymers, such as **9a,b**, is to use a TT–DPP–TT monomer. This would also allow the synthesis of a polymer with a greater number of TT units than would be accessible using the standard thiophene-flanked DPP monomer. In dilute chloroform solution, both polymers **9a** and **9b** display absorption maxima at the wavelength greater than 800 nm with a shoulder at a shorter wavelength [32b]. The spin-coated film of **9b** (chlorobenzene solution, 5 mg/mL) shows a significant redshift (~50 nm) in its absorption onset, which can be attributed to solid-state packing effect, whereas for **9a**, this increase is much smaller (~20 nm). However, the absorption maxima for both polymers are surprisingly blueshifted relative to the solution measurement, a result of what appears to be the shorter-wavelength shoulder increasing in intensity and becoming the dominant feature. Polymer **9b** was used in field-effect transistor devices and exhibited a maximum hole mobility of 1.95 cm^2 V^{-1} s^{-1}. Introducing the thienyl groups on the benzodithiophene (BDT) unit alters the LUMO and HOMO levels of the corresponding polymer **9c**, which has a fairly low absorption between 400 and 600 nm and absorbs the light intensely from 650 nm to 850 nm with a maximum near 800 nm. The absorption onset of **9c** is located at 858 nm, indicating an optical band gap of 1.44 eV. A single-layer photovoltaic device based on **9c** exhibited a power conversion efficiency of ~6%. When it was applied to tandem solar cells, a power conversion efficiency reached to 8.62% [36]. In 2012, the same research group further demonstrated a visibly transparent solar cell based on **9c** with a PCE of 4% and a maximum transparency of 66% at 550 nm in 2012 [37].

Porphyrin has been used as an electron-donating building block in making low band-gap conjugated polymers. Zhan et al. obtained three porphyrin-based polymers **10a-c** by the Sonogashira cross-coupling polymerization, containing the pyrido[3,4-*b*]pyrazine (**10a**) and perylene diimide (**10b**) and dithienothiophene (**10c**) segments [38]. Polymers **10a** and **10b** in films exhibit strong absorption in the NIR region (820–950 nm) with optical band gaps as low as 1.15 eV; their Q-bands redshift 60–190 nm compared to that of **10c**, while the Soret bands are similar. Regardless, the dithienothiophene unit in polymer **10c** is considered as an electron acceptor or donor, the absorption with a maximum at 761 nm can be assigned to a charge-transfer band.

	λ_{max}, nm (film)	λ_{max}, nm (solution)
10a	820	836
10b	948	912
10c	761	705

Perylene diimides (PDI) are extremely stable organic dyes that are actively explored for use in optoelectronic and photovoltaic devices, thermographic processes, energy-transfer cascades, light-emitting diodes, and NIR-absorbing systems [39]. The electron-accepting ability of PDI makes it a useful building block as an acceptor in construction of D–A conjugated polymers. The currently available synthesis allows coupling of PDI via its "bay" positions with other donor molecules to form conjugated polymers. The band gaps of the PDI-based polymers depend on the nature of donor units and also the twist angle or effective conjugation between the PDI and donor units, as represented by polymers **11a-f**. The presence of the planar dithienothiophene (DTT) units in polymer **11a** may benefit an ordered structure and offer a possibility of low-energy charge-transfer transition. A thin film of **11a** shows a broad absorption from 300 nm to 850 nm with the Q-band maximum at about 635 nm only and has a rather wide band gap of 1.9 eV [40]. The disappointing short-wavelength absorption maximum is mainly caused by a fairly large and variable twist between the two rigid PDI and DTT units, which hampers effective π-conjugation and also explains the observed long tailing of absorption into the NIR region. Insertion of a bridging unit such as acetylene can address this twist issue and enhance the D–A charge transfer, as evident by strong NIR absorption above 900 nm for polymer **10b**. As expected, simply inserting additional DTT units, the band gap (obtained from electrochemical data) could only decrease the band gap slightly to 1.7 eV (λ_{max} = 650 nm, chloroform) for polymer **11b** and 1.5 eV (λ_{max} = 690 nm, chloroform) for **11c** [41]. Using more electron-donating units of dithienopyrrole (DTP) and cyclopentadithiophene (CDT), a further reduction of band gap was achieved, as shown by polymers **11d** (1.6 eV, λ_{max} = 726 nm, chloroform) and **11e** (1.5 eV, λ_{max} = 715 nm, film) [42]. However, possibly due to a large twist again, polymer **11f** having the electron-rich benzodithiophene (BDT) unit was reported to have the band gap of 1.74 eV (λ_{max} = 549 nm, film) [43]. Furthermore, based on the same reasoning, conjugated D–A polymers derived from thiophene, bithiophene, fluorene, BDT, and other aromatic/heterocyclic molecules as donors and naphthalene diimide (NDI) as an acceptor show wide band gaps typically above 1.5 eV and the absorption maximum below 600 nm [43,44].

One exceptional case was reported for the NDI-based D–A polymers. Studies on a series of NDI-based conjugated polymers show that changing from thiophene donor to bithiophene donor (**12a**) causes a redshift of 125 nm in solution and 89 nm in annealed thin films [45]. Similar to PDI-based polymers **11**, polymer **12a** exhibited a broad absorption across the visible spectrum into the NIR with a maximum at 693 nm in THF and 707 nm in film. However, alkylation of the bithiophene units leads to a blueshift of 154–159 nm, while alkoxylation (**12b** versus **12a**) causes a very large redshift of nearly 300 nm. Electrostatic attraction between ether oxygen and thienyl sulfur atoms of **12b** is likely to enhance planarization [46], while mesomeric effects from pendant ether oxygen destabilize the HOMO, together causing the lowest optical band gap (1.08 eV). It is possible that inter- and intramolecular CT states contribute to the particularly low optical energy gap for polymer **12d**.

12a, R = H
λ_{max} = 693 nm (THF), 707 nm (film)

12b, R = OC$_{12}$H$_{25}$
λ_{max} = 985 nm (THF), 966 nm (film)

4.4   Donor–acceptor conjugated polymers

Engineering of a band structure of conjugated polymers affects greatly the optical and electronic properties of the polymers, resulting in their possible utilities for a variety of device applications, such as light-emitting diodes, photovoltaics, field-effect transistors, and photodetectors. In particular, combination of strong donors and strong acceptors can lead to materials simultaneously exhibiting low ionization potentials and high electron affinities that, in turn, can lead to the possibility of facile injection of both holes and electrons at moderate potentials and, therefore, the possibility of ambipolar charge transport. The NIR absorptions associated with the low band gap in the donor–acceptor conjugated polymers are attractive for organic solar cell, NIR photodetector, and other photonics applications. Accordingly, in the past 10 years, donor–acceptor conjugated polymers have attracted considerably more attention than nonconjugated polymers.

A large number of theoretical and experimental data indicate clearly that significant reduction of the band gap of conjugated polymers can be achieved by the application of three concepts: cancellation of bond-length alternation, introduction of the effective donor–acceptor interaction, and partial or complete planarization of conjugated polymer backbone. In addition, the band-gap engineering can be effective and quite significant at the macromolecular level, such as polymer interchain interaction, mesoscopic ordering, and intermolecular charge transfer by doping or blending. A common strategy in the polymer design involves the use of alternative donor and/or acceptor building blocks for constructing D–A polymers. To ensure effective conjugation length and efficient intermolecular charge transfer between the donor and acceptor units, introduction of the repeating D–π–A–π–D segments in the polymer main chain (Figure 4.3), in which π is a bridging unit or spacer, and planarization of part or ideally longer segments of the polymer backbone through fusing of donor or acceptor part with other aromatic rings or intramolecular bonding (e.g., hydrogen bonding and metal complexing) are preferred. Introduction of the quinoidal characters in the polymer backbone can lead to a further band-gap reduction. Selection of suitable donors, acceptors, and spacers is critical in obtaining the D–A conjugated polymers with the desired band gaps, charge mobility, and other optoelectronic properties and has been subjected to intensive studies, in particular, for applications in organic solar cells [47].

A trend in the design for the conjugated polymers with very low band gaps points to the use of strong electron acceptors and planar, fused thiophene molecules as electron donors, together with a suitable spacer. It is also clear that finding electron-accepting subunits with large atom orbital coefficients on the coupling units represents a crucial issue in designing donor–acceptor conjugated polymers with a low band gap. The degree of dispersion of the HOMO and LUMO levels depends on the size of the atom orbital coefficients on the coupling positions of the monomers. Therefore, it must be kept in mind that, when designing low band-gap D–A conjugated polymers, initial reduction of the energy separation between HOMO and LUMO levels cannot be scaled to homopolymers, and preferentially the acceptor unit should have its electron-accepting part incorporated in the conjugated backbone.

A π-spacer is often linked to an acceptor as a monomer used in polymerization and extends the overall conjugations in polymers. Common spacers include vinylene, ethynylene, phenylene, naphthylene, thiophene, and pyrrole (Figure 4.3). Phenylene and naphthylene are bulkier than the rest and less often used. Thiophene is considered aromatic, although theoretical calculations suggest that the degree of aromaticity is less than that of benzene. The general observation is an increase in resonance energy, either from heats of combustion and hydrogenation or from molecular orbital calculation, in the sequence of furan < pyrrole < thiophene < benzene.

The most obvious approach to the design of electron acceptors is to select an aryl or heteroaryl unit that bears one or more electronegative atoms in the ring, close to the coupling positions. Some of typical electron acceptors are presented in Figure 4.3. Based on the core structures, there are four series of acceptors that are known to date, namely, benzo, naphtho, thieno, and heterocyclic series. The benzo series represent 10π- or 14π-electron-heterocycles as *o*-quinoid acceptors derived from 1,2-phenlyenediamine. Cyclization using appropriate reagents yields the corresponding benzothiadiazole (BT), benzooxadiazole (BO), benzotriazole (BTA), and Se and

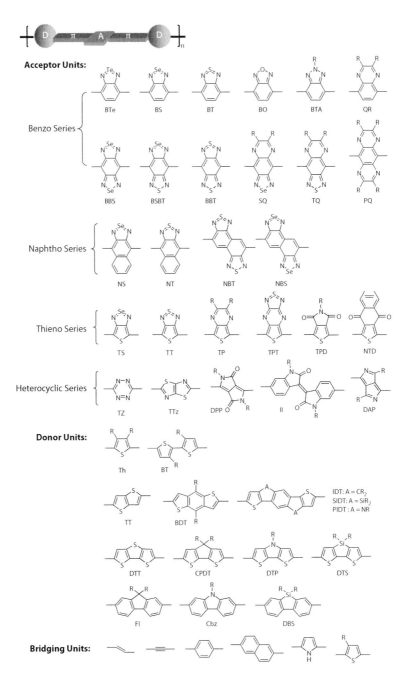

Figure 4.3 General structure of conjugated polymers having the repeating D–π–A–π–D segments and typical examples of electron-accepting, electron-donating, and bridging units useful for constructing the D–π–A–π–D conjugated polymers.

Te analogs (BS and BTe) [48]. Condensation of 1,2-phenylenediamine with 1,2-diketone affords the R-substituted quinoxaline (QR) acceptor. In this series, the acceptor strength increases in order of BTA, QR, BO, BT, BS, and BTe. Fusing another heterocyclic ring onto the above benzo-series acceptors give rise to an additional set of more powerful acceptors, with thiadiazolo[3,4-*g*]quinoxaline (TQ) being a common acceptor and Se analog (BBS) being the strongest electron-accepting one within this benzo series of acceptors. The benzo[1,2-*c*;3,4-*c'*]bis[1,2,5]-thiadiazole (BBT)-type unit is known to possess a substantial quinoidal character within a conjugated backbone, allowing for greater electron delocalization and thus lowering the band gap [49]. Fundamentally, the electron-accepting ability is governed by the LUMO energy level of the unit; the lower the LUMO energy level is, the stronger the electron-accepting ability. Since the structures of *o*-quinoid heterocycles affect primarily the LUMO levels, BBT has the LUMO of −3.21 eV, which is much lower than those of TQ (−2.39 eV) and pyrazino[2,3-*g*]quinoxaline (PQ) (−1.60 eV). Thus, BBT is more favorable than TQ and PQ for making the HOMO–LUMO separation small. Furthermore, the two selenium analogs, BBS and BSBT, are known to lead to further band-gap reduction [50].

By analogy, the naphtho series of acceptors are in principle derived from diaminonaphthalene and tatraaminonaphthalene, such as 2,1,3-naphthoselenadiazole (NS) and 2,1,3-naphthothiadiazole (NT) from 2,3-diaminonaphthalene. Naphtho[1,2-*c*:5,6-*c*]bis[1,2,5]thiadiazole (NBT) was first reported by a Japanese group in 1991 [51], and recently used for the first time as an acceptor in D–A conjugated polymers for solar cell application by the Cao group [52]. Due to extended conjugation in the quinoid form, NBT and its hypothetical Se analog (NBS) acceptors are expected to be more electron accepting than the corresponding benzo analogs although being less synthetically available.

The thieno series of acceptors are derived from 2,3-diaminothiophene precursors, such as thieno[3,4-*c*]thiadiazole (TT) and thieno[3,4-*b*]pyrazine (TP) acceptors, except for thieno[3,4-*c*]pyrrole-4,6-dione (TPD) naphthathiophenedione (NTD). The Se analogous acceptor (TS) is expected to be more electron accepting than TT and TP but is rather difficult to prepare. Compared to QR (−0.90 eV), TP has a lower LUMO level (−1.41 eV) and thus can lead to a lower band gap in the D–A conjugated polymers. The thieno-pyrazine-thiadiazole (TPT) is a promising acceptor that can lead to very low band-gap polymers, as a small molecule of thiophene-TPT-thiophene already shows an absorption peak at 990 nm. However, TPT and related electrochemically polymerized polymers were reported in only a few papers [53]. Thieno[3,4-*c*]pyrrole-4,6-dione (TPD) is relatively simple, compact, symmetric, and planar and could be beneficial for electron delocalization when it is incorporated into D–A conjugated polymers [54]. In all the thieno-series acceptors, the coupling positions are part of a

5-membered ring and flanked by an electron-donating sulfur atom, which is beneficial for the size of the atom orbital coefficients [55].

In the last series of acceptors, they all contain the electron-deficient heterocyclic rings. Due to the presence of electron-withdrawing imine group, s-tetrazine (Tz) [56] and thiazolothiazole (TTz) [57] are highly electron deficient; whereas the indigo-like, cross-conjugated vinylogous carbonyl units are responsible for the electron-accepting ability of diketopyrrolopyrrole (DPP) [58] and isoindigo (II) [59]. Finally, by comparing the LUMO level of 2,5-diazapentalene (DAP) (−3.70 eV) with that of a DPP compound (−3.13 eV), the former is a stronger acceptor. Accordingly, a DAP-based polymer has a lower band gap than the DPP counterpart [60].

The design of electron donors for use in conjugated polymers is primarily based on the two fundamental building blocks, thiophene and benzene. As shown in Figure 4.3, there are mainly four types of donor units: (i) linear substituted thiophene (Th) and bithiophene (DT), (ii) fused thiophene/benzene units, (iii) bridged dithiophene units, and (iv) bridged biphenyl units. To ensure a high electron-donating ability, the substituents on Th and DT are usually the alkoxy and alkyl groups, such as 3,4-ethylenedioxythiophene (EDOT). By fusing two thiophene rings or thiophene with benzene, a variety of fused donors are made available, for example, thieno[3,4-*b*]thiophene (TT), benzodithiophene (BDT), indacenodithiophene (IDT), silaindacenpdithiophene (SIDT), and pyrroloindacenodithiophene (PIDT). Another family of donor units is derived from dithiophene by bridging with S, C, Si, and N atom, as represented by dithienothiophene (DTT), cyclopentadithiophene (CPDT), dithienosilole (DTS), and dithienopyrrole (DTP), respectively. Similarly, the fluorene (Fl), carbazole (Cbz), and dibenzosilole (DBS) donors can be regarded as a biphenyl derivative bridged by a carbon, nitrogen, and silicon atom, respectively. By locking the dithiophene and biphenyl in one plane, the tricyclic system is much more efficient for electron delocalization than the parent units in the conjugated polymers. The bridging atoms also play a considerably positive role in adjusting the electronic property of the donors and as well polymers. Compared to the bridged biphenyl donors, the bridged dithiophene units have stronger orbital mixings with the acceptor units and tend to produce D–A polymers with a higher degree of planarity, as the flanking five-numbered thiophene rings cause less steric hindrance with their neighbor blocks than six-numbered rings do [61].

4.4.1   *Effect of spacer units*

The primary role that a spacer plays in the D–π–A–π–D type of conjugated polymers is to decrease the dihedral or torsion angle between two donor and acceptor units. It can make a significant difference in the band gap or absorption for conjugated polymers with or without a spacer between

Table 4.1 Optical and band-gap properties

	λ_{max}/nm (film)	λ_{max}/nm (solution)	Eg/eV (optical)	Eg/eV (electrochemical)
PQRV	568	558	1.86	1.99
PQRE	398	395	2.00	2.06
PTPV	1228	1211	0.88	1.00
PTPE	1220	1210	1.90	1.06

the bulky donor and acceptor, as shown by polymers **10b** (λ_{max} = 912 nm, with acetylene spacer) and **11d** (λ_{max} = 726 nm, without a spacer) having the same large bulky acceptor and the similar electron-rich donors. Chen et al. prepared a series of conjugated polymers, **PQRV**, **PQRE**, **PTPV**, and **PTPE**, consisting of QR and TP acceptor units that are connected by vinyl-ene and ethynylene spacers (Table 4.1) [62]. The optical band gaps of **PQRV**, **PQRE**, **PTPV**, and **PTPE** are 1.86, 2.00, 0.88, and 0.90 eV, respectively, whereas the electrochemical band gaps are 1.99, 2.06, 1.00, and 1.06 eV, respectively. The absorption maxima of **PQRV** are observed at 556 and 568 nm in chloroform solution and solid thin film, respectively, whereas those of **PQRE** are around 395 (with a shoulder peak around 493 nm) and 398 nm (with shoulder peak around 501 nm). It clearly indicates that the effect of spacer, vinylene versus ethynylene, on the π-π^* delocalization or band gap of QR-based two polymers. In addition, both QR-based polymers show much smaller optical band gap than their parent poly(quinoxaline)s (~2.43 eV) [63], due to the reduced steric hindrance between the QR accep-tor unit and vinylene or ethynylene spacer. The absorption spectra of **PTPV** and **PTPE** films show absorption maxima at 1228 and 1220 nm, respectively. Notably, the absorption bands of **PTPV** and **PTPE** are sig-nificantly redshifted when compared with their parent poly(thieno[3,4-*b*] pyrazine) without any spacer (~950 nm, 0.95 eV) that were synthesized by Grignard metathesis and FeCl$_3$ oxidative polymerizations [64]. It should be noted that the electrochemical polymerization of the same monomer yields poly(thieno[3,4-*b*]pyrazine) with a much lower band gap (0.66 eV) and long absorption (1275 nm) [65], which could not be satisfactorily explained without completely ruling out a possible electrochemical dop-ing, given that there is a fairly large twist between the repeating TP units in poly(thieno[3,4-*b*]pyrazine). The smaller band gap of **PTPV** (or **PTPE**) than that of **PQRV** is resulted from the stronger acceptor strength and more planar backbone of the former. Similar to the QR series, **PTPV** also exhibits stronger main-chain π-π^* transition than **PTPE**. There are some reports on this. One indicated that the ethynylene-linkage polymer had a larger torsional angle than vinylene-linkage one [66]. Therefore, the back-bone planarity of the vinylene-linked polymers could be better than that of the ethynylene-linked ones, which can lead to extended π-conjugation and smaller bond length alternation.

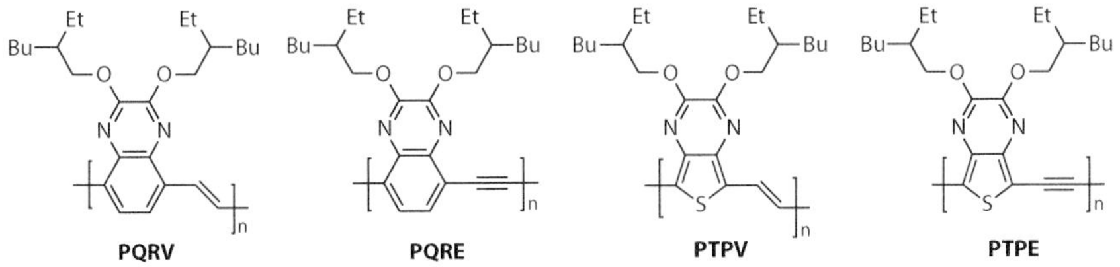

| PQRV | PQRE | PTPV | PTPE |

The thiophene spacer is widely used in D–A conjugated polymers for a number of reasons. The molecule is flat; the bond angle at the sulfur is around 93 degrees, the C-C-S angle is around 109, and the other two carbons have a bond angle around 114 degrees. Besides its small size and good thermal and photochemical stability, the thiophene spacer offers a synthetic viability as it can be directly polymerized electrochemically or readily converted to the bromide, boronic acid, Sn salt, and other reactive groups suitable for polycondensation. Furthermore, the thiophene spacer, when being connected to an acceptor that is able to form a quinoid structure such as BT, TT, TP, and BBT, is believed to be able to stabilize the quinoid form and thus lower the band gap (Figure 4.3). Anchoring of an alkyl chain on the thiophene spacer can impart a better solubility to the resulting polymer, but the alkyl chain position also affects the planarity. It has been found that substitution on the thiophene spacer at the position close to the acceptor unit induces a higher steric hindrance than the position close to the donor unit [67].

Owing to the presence of a N-H proton, the pyrrole spacer and the neighboring quinoid acceptor (e.g., BT and BBT) are expected to be completely coplanar by virtue of hydrogen bonding (Figure 4.4), which should facilitate the charge transfer from donor to acceptor and thus lowers the

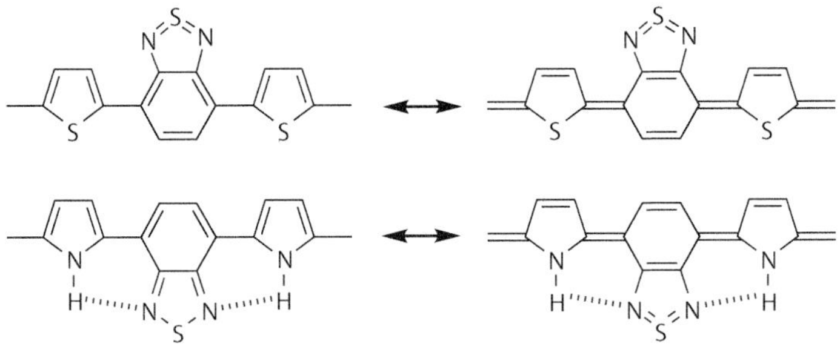

Figure 4.4 Band-gap reduction due to stabilization of the quinoid structure of the -π–A–π- moiety by the thiophene and pyrrole spacers and planarization due to intramolecular hydrogen bonding between BT acceptor and pyrrole spacer.

band-gap level. This intramolecular hydrogen bonding also contributes to altering the energy level of the quinoid acceptor by sharing or removing partial electrons on the nitrogen in acceptor. The presence of a hydrogen bond between the pyrrole N-H and the imine nitrogen of the BT or BBT unit has been confirmed by NMR (N-H peak at the low field, e.g., 11–12 ppm) and x-ray crystallographic studies on the model compounds [68].

4.4.2 Effect of electron acceptors

Given a large number of available acceptors, it is attempting to quantify the effect of each acceptor on the band gap of D–A conjugated polymers. One of the challenges with systematically determining the relationships between structure and band-gap property in D–A polymers is that fairly large functional groups are used to change and control electronic properties. A far more direct means to understand structure–property relationships in D–A polymers would be to carry out a study whereby a single atom is systematically varied in either the donor or acceptor moiety and properties are determined. Going from BT to BS to BTe acceptors, the structural variation occurs at the single atom, namely, from S to Se to Te, which would allow for pinpoint probing the changes in band gaps and absorptions of their structurally related polymers. Seferos et al. synthesized **PCPDTBT**, **PCPDTBS,** and **PCPDTBTe** polymers, wherein a single atom in the benzochalcogenodiazole unit is varied from S to Se to Te, and explicitly studied the effect of acceptors on the polymer properties [69]. They found that heavy atom substitution leads to a redshift in the low-energy transition, while the high-energy band remains relatively constant in energy. The band-gap values (and absorption maximum) obtained from the low-energy transition are 1.59 (700 nm), 1.46 (750 nm), and 1.06 eV (800 nm) for the S-, Se-, and Te-containing polymers, respectively. The present donor–acceptor theory predicates that optical properties are governed exclusively by the strength of D and A units. However, their experimental data reveal that the redshift in low-energy absorption is likely due to both a decrease in ionization potential and an increase in bond length and decrease in acceptor aromaticity. The loss of intensity of the low-energy band is likely the result of a loss of electronegativity and the acceptor unit's ability to separate charge. This work, therefore, provides a new way to control and understand the absorption properties of D–A polymers.

A similar atomistic band-gap engineering in D–A polymers was done almost at the same time by the Zade group, using a series of polymers derived from a 1,5-dithienyl-cyclopenta[*c*]thiophene unit as a donor and three different acceptors (BS, BT, and BTA) [70]. These polymers showed an interesting trend of visible color (red, green, and blue) in solution as the acceptor varied, while the donor remained the same. The optical band gaps of the polymer films were found to decrease from 1.86 to 1.57 to 1.44 eV

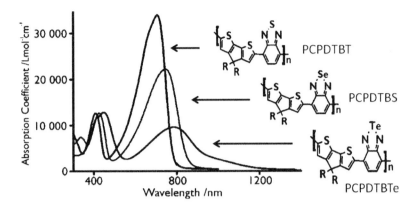

Figure 4.5 Absorption spectra of **PCPDTBT, PCPDTBS,** and **PCPDTBTe** in chloroform. (Adapted with permission from G. L. Gibson et al., *J. Am. Chem. Soc.* **2012**, *134*, 539. Copyright (2012) American Chemical Society.)

as the acceptor strength increases from BTA to BT to BS, respectively. The BS-based polymers have almost the identical band gaps (1.44 and 1.46 eV) as **PCPDTBS** (Figure 4.5), indirectly implying the two different donors (terthiophene and bridged dithiophene) have a similar electron-donating effect on the band gap.

The above studies further imply that certain types of acceptors can affect the band gaps of D–A polymers more than the donors. Among all the acceptors (Figure 4.3), BBS, BSBT, BBT, NBT, NBS, TS, TT, and TPT are the very strong acceptors and should be considered for designing D–A polymers having band gaps below 1.0 eV or absorptions over 1000 nm. For example, a D–π–A–π–D polymer **(PBBTDTP)** derived from BBT (A), DTP (D), and thiophene (π) displays the lowest-energy absorption maximum at 1154 nm (1160 nm as film), which is equivalent to an optical band gap of 0.56 eV [71]. By changing the donor unit of DTP to thieno[3,2-*b*] thiophene (TT) unit, the resulting polymer **PBBTTT** absorbs in the NIR region with the absorption onset of around 2200 nm in the solid state, which corresponds to an optical band gap of 0.56 eV (Figure 4.6) [72]. This is probably the most well-characterized, chemically synthesized organic polymer with the lowest band gap and the longest absorption wavelength. Furthermore, it is interesting to note that the absorption spectra of **PBBTTT** in solution and in film are almost identical, indicating no or little interchain stacking in the solid state. Finally, using even a weak donor of terthiophene to connect the thieno[3,4-*c*]thiadiazole acceptor unit, the resulting structurally rather simple polymer **PTTTT** was reported to have the lowest absorption band ranging from 600 to 1200 nm with a maximum at about 850 nm [73].

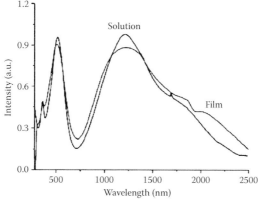

Figure 4.6 NIR-absorbing polymers: **PBBTDPT**, **PBBTTT**, and **PTTT**. Absorption spectra of **PBBTTT** in chloroform and as thin film. (Adapted with permission from J. Fan et al., *Adv. Mater.* **2012**, 24, 2186.)

Aiming at achieving a magical band-gap value of 0 eV, two monomers (**DT-TPT** and **DP-TPT**) were prepared by end-capping TPT with two thiophene and two pyrrole units and then electrochemically polymerized. The thiophene-containing polymer **PTPTDT**, after adequate de-doping, showed a very low band gap of 0.3 eV [53a]. Monomer **DP-TPT** was prepared by oxidation with NiO₂ and, due to its high reactivity, could not be isolated [53b]. However, since the oxidation process was very clean, a solution of pure **DP-TPT** could be used directly in electrochemical polymerization after removal of the nickel salts by filtration. The lowest-energy absorption maximum and the absorption edge of this monomer in solution were observed at 1345 nm (0.92 eV) and 2100 nm (~0.6 eV), respectively, which is an extremely high value for such a small organic molecule.

The anodic oxidation gave the electroactive polymer **PTPTDP** exhibiting a vanishingly small electrochemical energy gap. Electrochemical determination of the band gap was not possible, since there was no potential gap between n- and p-doping peaks. From this fact, it was concluded that the band gap of **PTPTDP** is zero. Unfortunately, no optical data were presented to support this hypothesis. Furthermore, it was stated that de-doping of this electrochemically prepared polymer was not possible due to the zero band gap, which leaves the possibility of residual doping open.

4.4.3   Effect of electron donors

The effect of electron donors on the band gaps or absorptions of D–A polymers is mainly related to the energy gap and structural features of donor molecules. The small energy gap and/or high-lying HOMO level, fused ring structure, and extended conjugation are all in favor of narrowing the band gap of the corresponding D–A polymers. Among all the listed donor units in Figure 4.3, the benzene and biphenyl-based donors are generally weaker than the thiophene and dithiophene-based donors for a number of fundamental reasons. In comparison, benzene is more aromatic but less electron-donating than thiophene and biphenyl has a larger dihedral angle or less coplanar than dithiophene. Accordingly, the TT, BDT, PIDT, DTT, CPDT, DTP, and DTS donors are preferred for making low band-gap D–A polymers. Because flanking five-numbered thiophene rings cause less steric hindrance with the neighboring units in the polymer, these bridged dithiophene donors and even BT enable a better electron delocalization and thus stronger orbital mixings with acceptor units, which are in favor of producing D–A polymers with high planarity and effective conjugation.

PTQFl	**PTQCbz**	**PTQDTP**
λ_{max} (solution) = 840 nm	847 nm	931 nm

For example, three D–A polymers, **PTQFl** [74], **PTQCbz** [75], and **PTQDTP** [71], share the same acceptor unit of TQ and have different donor units. The polymers with the Fl and Cbz units have almost the identical absorption maxima at 840 and 847 nm, although Fl and Cbz have different electron-donating powers. However, with the dithiophene-based DTP donor unit, the absorption maximum of **PTQDTP** shifts nearly 90 nm bathochromically to 931 nm. This large redshift in absorption is mainly due to the difference in orbital mixings between the donor and acceptor units through adjacent thiophene spacer.

References

1. (a) A. Berlin, A. Canavesi, G. Pagani, G. Schiavon, S. Zecchin, G. Zotti, *Synth. Met.* **1997**, *84*, 451. (b) G. Zotti, S. Zecchin, G. Schiavon, A. Berlin, G. Pagani, M. Borgonovo, R. Lazzaroni, *Chem. Mater.* **1997**, *9*, 2876. (c) S.-C. Lin, J.-A. Chen, M.-H. Liu, Y. O. Su, M.-K. Leung, *J. Org. Chem.* **1998**, *63*, 5059.
2. (a) W. Kaim, V. Kasack, H. Binder, E. Roth, J. Jordanov, *Angew. Chem. Int. Ed. Engl.* **1988**, *27*, 1174. (b) W. Kaim, V. Kasack, *Inorg. Chem.* **1990**, *29*, 4696. (c) V. Kasack, W. Kaim, H. Binder, J. Jordanov, E. Roth, *Inorg. Chem.* **1995**, *34*, 1924. (d) S. Wang, X. Li, S. Xun, X. Wan, Z. Y. Wang, *Macromolecules*, **2006**, *39*, 7502.
3. Y. Qi, P. Desjardins, Z. Y. Wang, *J. Opt. A: Pure Appl. Opt.* **2002**, *4*, 1.
4. (a) J. R. Reynolds, F. E. Karasz, C. P. Lillya, J. C. W. Chien, *J. Chem. Soc. Chem. Commun.* **1985**, 268. (b) J. R. Reynolds, J. C. W. Chien, C. P. Lillya, *Macromolecules* **1987**, *20*, 1184. (c) J. R. Reynolds, C. A. Jollv, S. Krichene, P. Cassoux, C. Faulmann, *Synth. Met.* **1989**, *31*, 109.
5. (a) E. M. Enaler, K. H. Nichols, V. V. Patel, N. M. Rivera, R. R. Schumaker, U.S. Patent 4,111,857 (1978). (b) H. Poleschner, W. John, F. Hoppe, E. Fanghanel, J. *Prakt. Chem.* **1983**, *325*, 957. (c) R. Vicente, J. Ribas, P. Cassoux, L. Valade, *Synth. Met.* **1986**, *13*, 265.
6. F. Gotzfried, W. Beck, A. Lerf, A. Sebald, *Angew. Chem. Int. Ed. Engl.* **1979**, *18*, 463.
7. (a) N. M. Rivera, E. M. Engler, R. R. Schumaker, *J. Chem. Soc. Chem. Commun.* **1979**, 164. (b) J. Ribas, P. Cassoux, *C. R. Acad. Sci. Ser. 2*, **1981**, *293*, 665.
8. B. K. Teo, F. Wudl, J. J. Hauser, A. Krueer, *J. Am. Chem. Soc.* **1977**, *99*, 4862.
9. C. W. Dirk, M. Bosseau, P. H. Barrett, F. Moraes, F. Wudl, A. J. Heeger, *Macromolecules*, **1986**, *19*, 266.
10. (a) F. Wang, J. R. Reynolds, *Macromolecules*, **1988**, *21*, 2887. (b) F. Wang, J. R. Reynolds, *Macromolecules*, **1990**, *23*, 3219.
11. (a) I. Tabushi, K. Vamamura, H. Nonoguchi, *Chem. Lett.* **1987**, 1373. (b) D. C. Olson, V. P. Mayweg, G. N. Schrauzer, *J. Am. Chem. Soc.* **1966**, *88*, 4876.
12. A. Kobayashi, Y. Sasaki, H. Kobayashi, A. E. Underhill, M. M. Ahmad, *J. Chem. Soc. Chem. Commun.* **1982**, 390.
13. C. Ornelas, R. Pennell, L. F. Liebes, M. Weck, *Org. Lett.* **2011**, *13*, 976.
14. R. E. Peierls, *Quantum Theory of Solids*, Oxford University Press: London, 1956.
15. F. Wudl, N. Kobayashi, A. J. Heeger, *J. Org. Chem.* **1984**, *49*, 3381.
16. J. L. Brédas, A. J. Heeger, F. Wudl, *J. Chem. Phys.* **1986**, *85*, 4673.
17. Y. Ikenoue, F. Wudl, A. J. Heeger, *Synth. Met.* **1991**, *40*, 1.
18. M. Pomerantz, B. Chaloner-Gill, L. O. Harding, J. J. Tseng, W. J. Pomerantz, *Synth. Met.* **1993**, *55–57*, 960.
19. (a) C. Arbizzani, M. Catellani, M. Grazia Cerroni, M. Mastragostoni, *Synth. Met.* **1997**, *84*, 249. (b) M. Pomerantz, X. Gu, *Synth. Met.* **1997**, *84*, 243. (c) S. Inaoka, D. M. Collard, *Synth. Met.* **1997**, *84*, 193.
20. L. Wen, B. C. Duck, P. C. Dastoor, C. Rasmussen, *Macromolecules*, **2008**, *41*, 4576.
21. M. Pomerantz, X. Gu, S. X. Zhang, *Macromolecules*, **2001**, *34*, 1817.
22. H. Reisch, U. Wiesler, U. Scherf, N. Tuytuylkov, *Macromolecules*, **1996**, *29*, 8204.
23. (a) G. Brocks, A. Tol, *Synth. Met.* **1996**, *76*, 213. (b) G. Brocks, A. Tol, *J. Phys. Chem.* **1996**, *100*, 1838.

24. (a) A. Treibs, K. Jacob, *Angew. Chem. Int. Ed. Engl.* **1965**, *4*, 694. (b) Y.-Y. Chen, H. K. Hall Jr. *Polym. Bull.* **1986**, *16*, 419.

25. (a) E. E. Havinga, W. ten Hoeve, H. Wynberg, *Polym. Bull.* **1992**, *29*, 119. (b) E. E. Havinga, W. ten Hoeve, H. Wynberg, *Synth. Met.* **1993**, *55*, 299. (c) E. E. Havinga, A. Pomp, W. ten Hoeve, H. Wynberg, *Synth. Met.* **1995**, *69*, 581.

26. (a) D. E. Lynch, U. Geissler, J. Kwiatkowski, A. K. Whittaker, *Polym. Bull.* **1997**, *38*, 493. (b) D. E. Lynch, U. Geissler, I. R. Peterson, M. Floersheimer, R. Terbrack, L. F. Chi, H. Fuchs, N. J. Calos, B. Wood, C. H. L. Kennard, G. J. Langley, *J. Chem. Soc. Perkin Trans. 2* **1997**, 827. (c) D. E. Lynch, Floersheimer, D. Essing, L. F. Chi, H. Fuchs, N. J. Calos, B. Wood, C. H. L. Kennard, G. J. Langley, *J. Chem. Soc. Perkin Trans. 2* **1998**, 779.

27. (a) A. Ajayaghosh, C. R. Chenthamarakshan, S. Das, M. V. George, *Chem. Mater.* **1997**, *9*, 644. (b) C. R. Chenthamarakshan, A. Ajayaghosh, *Chem. Mater.* **1998**, *10*, 1657. (c) C. R. Chenthamarakshan, J. Eldo, A. Ajayaghosh, *Macromolecules*, **1999**, *32*, 251.

28. (a) A. Ajayaghosh, *Acc. Chem. Res.* **2005**, *38*, 449. (b) E. Arunkumar, A. Ajayaghosh, J. Daub, *J. Am. Chem. Soc.* **2005**, *127*, 3156. (c) S. Sreejith, P. Carol, P. Chithra, A. Ajayaghosh, *J. Mater. Chem.* **2008**, *18*, 264. (d) M.-C. Ho, C.-H. Chao, C.-H. Chen, R.-J. Wu, W.-T. Whang, *Macromolecules*, **2012**, *45*, 3010.

29. (a) J. Eldo, E. Arunkumar, A. Ajayaghosh, *Tetrahedron Lett.* **2000**, *41*, 6241. (b) A. Ajayaghosh, J. Eldo, *Org. Lett.* **2001**, *3*, 2595. (c) J. Eldo, A, Ajayaghosh, *Chem. Mater.* **2002**, *14*, 410.

30. Z. Hao, A. Iqbal, *Chem. Soc. Rev.* **1997**, *26*, 203.

31. (a) L. Bürgi, M. Turbiez, R. Pfeiffer, F. Bienewald, H.-J. Kirner, C. Winnewisser, *Adv. Mater.* **2008**, *20*, 2217. (b) D. Cao, Q. Liu, W. Zeng, S. Han, J. Peng, S. Liu, *Macromolecules*, **2006**, *39*, 8347.

32. (a) J. C. Bijleveld, A. P. Zoombelt, S. G. J. Mathijssen, M. M. Wienk, M. Turbiez, D. M. de Leeuw, R. A. J. Janssen, *J. Am. Chem. Soc.* **2009**, *131*, 16616. (b) H. Bronstein, Z. Chen, R. S. Ashraf, W. Zhang, J. Du, J. R. Durrant, P. Shakya Tuladhar, K. Song, S. E. Watkins, Y. Geerts, M. M. Wienk, R. A. J. Janssen, T. Anthopoulos, H. Sirringhaus, M. Heeney, I. McCulloch, *J. Am. Chem. Soc.* **2011**, *133*, 3272. (c) J. C. Bijleveld, V. S. Gevaerts, D. Di Nuzzo, M. Turbiez, S. G. J. Mathijssen, D. M. de Leeuw, M. M. Wienk, R. A. J. Janssen, *Adv. Mater.* **2010**, *22*, E242.

33. (a) Y. Li, S. P. Singh, P. Sonar, *Adv. Mater.* **2010**, *22*, 4862. (b) P. Sonar, S. P. Singh, Y. Li, M. S. Soh, A. Dodabalapur, *Adv. Mater.* **2010**, *22*, 5409. (c) Y. Li, P. Sonar, S. P. Singh, M. S. Soh, M. van Meurs, J. Tan, *J. Am. Chem. Soc.* **2011**, *133*, 2198. (d) P. Sonar, S. P. Singh, Y. Li, Z.-E. Ooi, T.-j. Ha, I. Wong, M. S. Soh, A. Dodabalapur, *Energy Environ. Sci.* **2011**, *4*, 2288.

34. (a) G. Y. Chen, C. M. Chiang, D. Kekuda, S. C. Lan, C. W. Chu, K. H. Wei, *J. Polym. Sci. Part A*, **2010**, *48*, 1669. (b) E. Zhou, Q. Wei, S. Yamakawa, Y. Zhang, K. Tajima, C. Yang K. Hashimoto, *Macromolecules*, **2010**, *43*, 821. (c) E. Zhou, S. Yamakawa, K. Tajima, C. Yang, K. Hashimoto, *Chem. Mater.* **2009**, *21*, 4055.

35. I. McCulloch, M. Heeney, C. Bailey, K. Genevicius, I. MacDonald, M. Shkunov, D. Sparrowe, S. Tierney, R. Wagner, W. Zhang, M. L. Chabinyc, R. J. Kline, M. D. McGehee, M. F. Toney, *Nature Mater.* **2006**, *5*, 328.

36. L. Dou, J. B. You, J. Yang, C. C. Chen, Y. J. He, S. Murase, T. Moriarty, K. Emery, G. Li, Y. Yang, *Nature Photonics*, **2012**, *6*, 180.
37. C.-C. Chen, L. Dou, R. Zhu, C.-H. Chung, T.-B. Song, Y. B. Zheng, S. Hawks, G. Li, P. S. Weiss, Y. Yang, *ACS Nano*, **2012**, *6*, 7185.
38. X. B. Huang, Q. Q. Shi, W. Q. Chen, C. L. Zhu, W. Y. Zhou, Z. Zhao, X. M. Duan, X. W. Zhan, *Macromolecules*, **2010**, *43*, 9620.
39. T.Weil, T. Vosch, J. Hofkens, K. Peneva, K. Müllen, *Angew. Chem. Int. Ed.* **2010**, *49*, 9068.
40. X. Zhan, Z. A. Tan, B. Domercq, Z. An, X. Zhang, S. Barlow, Y. Li, D. Zhu, B. Kippelen, S. R. Marder, *J. Am. Chem. Soc.* **2007**, *129*, 7246.
41. X. Zhan, Z. A. Tan, E. Zhou, Y. Li, R. Misra, A. Grant, B. Domercq, X.-H. Zhang, Z. An, X. Zhang, S. Barlow, B. Kippelen, S. R. Marder, *J. Mater. Chem.* **2009**, *19*, 5794.
42. (a) J. Hou, S. Zhang, T. L. Chen, Y. Yang, *Chem. Commun.* **2008**, 6034. (b) E. Zhou, J. Cong, Q. Wei, K. Tajima, C. Yang, K. Hashimoto, *Angew. Chem. Int. Ed.* **2011**, *123*, 2851.
43. J. Chen, M. M. Shi, X.-L. Hu, M. Wang, H.-Z. Chen, *Polymer* **2010**, *51*, 2897.
44. P. Piyakulawat, A. Keawprajak, A. Chindaduang, A. Helfer, U. Asawapirom, *e-Polymers*, **2010**, 071.
45. X. G. Guo, M. D. Watson, *Org. Lett.* **2008**, *10*, 5333.
46. (a) J. A. Irvin, I. Schwendeman, Y. Lee, K. A. Abboud, J. R. Reynolds, *J. Polym. Sci. A. Polym. Chem.* **2001**, *39*, 2164. (b) M. Pomerantz, *Tetrahedron Lett.* **2003**, *44*, 1563.
47. (a) Y. Li, *Acc. Chem. Res.* **2012**, *45*, 723. (b) Z.-G. Zhang, *J. Mater. Chem.* 2012, 22, 4178.
48. (a) J. Bouffard, T. M. Swager, *Macromolecules*, **2008**, *41*, 5559. (b) P. Ding, Y. Zou, H. Wu *J. Phys. Chem. C*, **2011**, *115*, 16211. (c) J.-M. Jiang, K.-H. Wei *Macromolecules*, **2011**, *44*, 9155. (d) J. Min, Z. Zhang, S. Zhang, M. Zhang, J. Zhang, Y. Li, *Macromolecules*, **2011**, *44*, 7632.
49. (a) K. Ono, S. Tanaka, Y. Yamashita, *Angew. Chem., Int. Ed. Engl.* **1994**, *33*, 1977. (b) M. Karikomi, C. Kitamura, S. Tanaka, Y. Yamashita, *J. Am. Chem. Soc.* **1995**, *117*, 6791. (c) C. Kitamura, S. Tanaka, Y. Yamashita, *Chem. Mater.* **1996**, *8*, 570. (d) Y. Yamashita, K. Ono, M. Tomura, S. Tanaka, *Tetrahedron* **1997**, *53*, 10169.
50. R. Yang, R. Tian, Q. Hou, W. Yang, Y. Cao, *Macromolecules*, **2003**, *36*, 7453.
51. M. Shuntaro, K. Takahashi, Y. Ikezaki, T. Hatta, A. Tori-i, M. Tashiro, *Bull. Chem. Soc. Jpn.* **1991**, *64*, 68.
52. M. Wang, X. Hu, P. Liu, W. Li, X. Gong, F. Huang, Y. Cao, *J. Am. Chem. Soc.* **2011**, *133*, 9638.
53. (a) S. Tanaka, Y. Yamashita, *Synth. Met.* **1995**, *69*, 599. (b) S. Tanaka, Y. Yamashita, *Synth. Met.* **1997**, *84*, 229.
54. (a) Y. Zou, A. Najari, P. Berrouard, S. Beaupre, B. Aich, Y. Tao, M. Leclerc, *J. Am. Chem. Soc.* **2010**, *132*, 5330. (b) T. Chu, J. Lu, S. Beaupr, Y. Zhang, J. Pouliot, S. Wakim, J. Zhou, M. Leclerc, Z. Li, J. Ding, Y. Tao, *J. Am. Chem. Soc.* **2011**, *133*, 4250.
55. A. K. Bakhshi, H. Ago, K. Yoshizawa, K. Tanaka, T. Yamabe, *J. Chem. Phys.* **1996**, *104*, 5528.
56. (a) Z. Li, J. Ding, N. Song, J. Lu, Y. Tao, *J. Am. Chem. Soc.* **2010**, *132*, 13160. (b) J. Ding, Z. Li, Z. Cui, G. P. Robertson, N. Song, X. Du, L. Scoles, *J. Polym. Sci. Part A: Polym. Chem.* **2011**, *49*, 3374.

57. (a) S. K. Lee, J. M. Cho, Y. Goo, W. S. Shin, J.-C. Lee, W.-H. Lee, I.-N. Kang, H.-K. Shim, S.-J. Moon, *Chem. Commun.* **2011**, *47*, 1791. (b) *Polym. Chem.* **2011**, *2*. (c) L. Huo, X. Guo, S. Zhang, Y. Li, J. Hou, *Macromolecules*, **2011**, *44*, 4035. (d) S. Subramaniyan, H. Xin, F. S. Kim, S. Shoaee, J. R. Durrant, S. A. Jenekhe, *Adv. Energy Mater.* **2011**, *1*, 854.

58. (a) E. Zhou, Q. Wei, S. Yamakawa, Y. Zhang, K. Tajima, C. Yang, K. Hashimoto, *Macromolecules*, **2009**, *43*, 821. (b) M. M. Wienk, M. Turbiez, J. Gilot and R. A. J. Janssen, *Adv. Mater.* **2008**, *20*, 2556.

59. (a) E. Wang, Z. Ma, Z. Zhang, K. Vandewal, P. Henriksson, O. Inganäs, F. Zhang, M. R. Andersson, *J. Am. Chem. Soc.* **2011**, *133*, 14244. (b) R. Stalder, C. Grand, J. Subbiah, F. So, J. R. Reynolds, *Polym. Chem.* **2012**, *3*, 89. (c) B. Liu, Y. Zou, B. Peng, B. Zhao, K. Huang, Y. He, C. Pan, *Polym. Chem.* **2011**, *2*, 1156.

60. (a) G. Qian, J. Qi, Z. Y. Wang, *J. Mater. Chem.* **2012**, *22*, 12867. (b) G. Qian, J. Qi, J. A. Davey, J. S. Wright, Z. Y. Wang, *Chem. Mater.* **2012**. *24*, 2364.

61. J. Ku, Y. Lansac, Y. H. Jang, *J. Phys. Chem. C*, **2011**, *115*, 21508.

62. C. C. Chueh, M. H. Lai, J. H. Tsai, C. F. Wang, W. C. Chen, *J. Polym. Sci. Part A*, **2010**, *48*, 74.

63. T. Yamamoto, K. Sugiyama, T. Kushida, T. Inoue, T. Kanbara, *J. Am. Chem. Soc.* **1996**, *118*, 3930.

64. (a) M. Pomerantz, B. Chalonergill, L. O. Harding, J. J. Tseng, W. J. Pomerantz, *J. Chem. Soc. Chem. Commun.* **1992**, 1672. (b) L. Wen, B. C. Duck, P. C. Dastoor, S. C. Rasmussen, *Macromolecules*, **2008**, *41*, 4576.

65. D. D. Kenning, S. C. Rasmussen, *Macromolecules*, **2003**, *36*, 6298.

66. (a) M. Levitus, K. Schmieder, H. Ricks, K. D. Shimizu, U. H. F. Bunz, M. A. Garcia-Garibay, *J. Am. Chem. Soc.* **2001**, *123*, 4259. (b) R. Pizzoferrato, M. Berliocchi, A. Di Carlo, P. Lugli, M. Venanzi, A. Micozzi, A. Ricci, C. Lo Sterzo, *Macromolecules*, **2003**, *36*, 2215. (c) D. L. Zeng, J. W. Chen, Z. Chen, W. H. Zhu, J. He, F. Yu, H. Huang, H. R. Wu, C. Liu, S. J. Ren, J. P. Du, J. Sun, E. J. Xu, A. Cao, Q. Fang, *Macromol. Rapid Commun.* **2007**, *28*, 772. (d) H. Fukumoto, T. Yamamoto, *J. Polym. Sci. Part A: Polym. Chem.* **2008**, *46*, 2975. (e) K. Tamura, M. Shiotsuki, N. Kobayashi, T. Masuda, F. Sanda, *J. Polym. Sci. Part A: Polym. Chem.* **2009**, *47*, 3506. (f) S. M. Zhang, H. J. Fan, Y. Liu, G. J. Zhao, Q. K. Li, Y. F.; Li, X. W. Zhan, *J. Polym. Sci. Part A: Polym. Chem.* **2009**, *47*, 2843. (g) J. G. Kushmerick, D. B. Holt, S. K. Pollack, M. A. Ratner, J. C. Yang, T. L. Schull, J. Naciri, M. H. Moore, R. Shashidhar, *J. Am. Chem. Soc.* **2002**, *124*, 10654.

67. (a) H. Zhou, L. Yang, S. Xiao, S. Liu, W. You, *Macromolecules*, **2009**, *43*, 811. (b) M. Zhang, X. Guo, Y. Li, *Adv. Energy Mater.* **2011**, *1*, 557.

68. (a) H. A. M. van Mullekom, J. A. J. M. Vekemans, E. W. Meijer, *Chem. Eur. J.* **1998**, 1235. (b) A. R. A. Palmans, J. A. J. M. Vekemans, E. W. Meijer, *Recl. Trav. Chim. Pays-Bas*, **1995**, *114*, 277. (c) A. R. A. Palmans, J. A. J. M. Vekemans, H. Fischer, R. A. Hikmet, E. W. Meijer, *Chem. Eur. J.* **1997**, *3*, 300. (d) G. Qian, Z. Y. Wang, *Can. J. Chem.* **2010**, *88*, 192.

69. G. L. Gibson, T. M. McCormick, D. S. Seferos, *J. Am. Chem. Soc.* **2012**, *134*, 539.

70. S. Das, P. B. Pati, S. S. Zade, *Macromolecules*, **2012**, *45*, 5410.

71. X. Zhang, T. T. Steckler, R. R. Dasari, S. Ohira, W. J. Potscavage Jr., S. P. Tiwari, S. Coppée, S. Ellinger, S. Barlow, J.-L. Brédas, B. Kippelen, J. R. Reynolds, S. R. Marder, *J. Mater. Chem.* **2010**, *20*, 123.

72. J. Fan, J. D. Yuen, M. Wang, J. Seifter, J.-H. Seo, A. R. Mohebbi, D. Zakhidov, A. Heeger, F. Wudl, *Adv. Mater.* **2012**, *24*, 2186.

73. Y. Xia, L. Wang, X. Deng, D. Li, X. Zhu, Y. Cao, *Appl. Phys. Lett.* **2006**, *89*, 081106.
74. M. X. Chen, X. Crispin, E. Perzon, M. R. Andersson, T. Pullerits, M. Andersson, O. Inganas, M. Berggren, *Appl. Phys. Lett.* **2005**, *87*, 252105.
75. H. Yi, R. G. Johnson, A. Iraqi, D. Mohamad, R. Royce, D. G. Lidzey, *Macromol. Rapid Commun.* **2008**, *29*, 1804.

chapter five

Emerging applications of near-infrared organic materials

Knowledge is a treasure, but practice is the key to it.

Thomas Fuller (1608–1661)

Most applications of NIR organic materials are being explored by utilizing directly and indirectly their unique properties in the NIR spectral region. The absorption, fluorescence, photosensitizing, photoconductive, photochromic, and other properties in the NIR spectral region are all being explored for various applications in many sectors. For example, NIR dyes are used as contrast-enhancing agents for spectroscopic optical coherence tomography [1] and as an NIR photosensitizer in photodynamic therapy for alternative treatment for cancer and several other medical conditions [2,3]. In this chapter, the discussion is focused on the applications that utilize the NIR-absorbing, NIR chromogenic (e.g., electrochromic), photosensitizing, photovoltaic, and fluorescent properties. In addition, applications of low band-gap compounds and polymers in ambipolar organic transistors are presented.

5.1 Applications based on NIR-absorbing property

Early reviews and reports have already given the detailed accounts for a number of applications and products, although some might be outdated or obsolete nowadays [4]. One of the most important properties of NIR materials that can be utilized for potential applications is the absorption at the specific NIR wavelength or conversion of the absorbed light energy into the heat. Some devices and products are made or operated by utilizing the heat generated from the NIR absorbers or dyes, such as the recording DRAW (Direct Reading after Writing) or WORM (Write Once Read Many) disk can operate in the heat-mode system using NIR organic dyes and the displays that are thermally addressed by a laser beam.

NIR absorbers, either colored or colorless to the naked eye, can find a wide range of applications, such as security printing (bank notes, credit cards, identity cards, and passports), invisible inks, invisible and NIR readable bar codes, laser welding and laser marking of plastics, thermal curing, optical filters for laser protective products, and plasma display panels [5]. Typical inorganic colorless NIR absorbers include lanthanum hexaboride, indium tin oxide, and antimony tin oxide. Sumitomo Metal Mining in Japan has developed lanthanum hexaboride nanoparticles for use as solar control coatings on PET window film. Lanthanum hexaboride is a semiconducting material and exhibits an intense absorption in the NIR when it is in the nanometer size regime. The absorption is due to a phenomenon called *surface plasmon resonance* that is a result of the nanoparticles' optical behavior being determined by its surface electrons, which are a higher proportion than its bulk electrons when the particles are in nanosizes. In collaboration with Sumitomo Metal Mining, a supplier of automotive glass interlayer products, Solutia, has debuted a new product—Vanceva Solar—that uses lanthanum hexaboride nanoparticles in polyvinyl butyral (PVB) to accomplish NIR absorption. The first use of the Vanceva Solar product was demonstrated with the Citroën C4 automobile in its panoramic glass roof to provide similar benefits associated with established technologies, such as tinted glass and coated glass, for solar control glazing (Figure 5.1). The success can be attributed to its excellent IR-blocking performance, long-term durability to UV exposure, and compatibility with typical PVB processing methods and performance properties, such as adhesion to glass, impact strength, and optical clarity. The product is essentially identical to reflective coating on glass or PET film in terms of heat load gain when the vehicle is moving.

In comparison with inorganic colorless NIR absorbers, most of the organic NIR dyes have absorption in the visible spectrum from 400 to 700 nm and suffer from poor UV durability and thermal degradation. Thus, improving the UV photostability and thermal stability is a key to success in the development of organic NIR absorbers. Several classes

Figure 5.1 The Citroën C4, introduced at the 2004 Paris Auto Show, features a panoramic laminated glass roof with solar-absorbing LaB_6 nanoparticles embedded in the PVB interlayer.

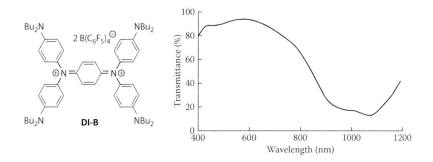

Figure 5.2 Vis–NIR spectrum of diimmonium borate **DI-B** (R^1-R^{10} = n-Bu).

of organic NIR dyes that may qualify as NIR absorbers are known, for example, polymethines, phthalocyanines, naphthacyanines, metal dithiolene complexes, and diimmonium dyes. Among them, a diimmonium dye has been explored for use in a plasma display panel (PDP) as an optical filter that can absorb the unwanted infrared rays radiated from the display, due to its high NIR light absorption at a wavelength above 900 nm and high transparency in the visible light region [6]. PDP generates a NIR ray of 800–1000 nm wavelengths on plasma discharge, and it has been a problem that this NIR ray induces improper operation of a remote controller of home electric appliances. In addition, because an optical semiconductor element used in a CCD camera has a high sensitivity in the NIR region, the removal of NIR ray is required. However, NIR-absorbing diimmonium dyes tend to deteriorate over time, resulting in change in NIR absorbance and coloration. This deterioration is deemed to be brought about by degeneration of a dye, caused by various factors such as heat, moisture, and light. A patent claims a series of NIR-absorbing diimmonium dyes (**DI-B**, Figure 5.2) having tetrakis(pentafluorophenyl) borate as a counter anion, which enables to improve durability, in particular, heat resistance and moisture resistance [6b]. For example, one of **DI-B** dyes absorbs around 1000 nm and shows good transparency in the visible spectral region (Figure 5.2). After heating at 120°C for 120 h, this dye showed only a very minor change in the NIR absorbance in comparison with the same dye having SbF_6 as a counter anion.

Quaterrylene diimides have a minimum of inherent colors and are shown to be extremely photostable and temperature stable, weather resistant, chemically inert, and environmentally compatible, making them an ideal heat protection additive for numerous transparent applications in the architectural and automobile sectors. In 2004, BASF announced the development of the transparent interlayer films containing quaterrylene diimide dyes as NIR absorbers for applications in sandwich-structured safety glazing in buildings, car windows, translucent roofs, greenhouses, etc.

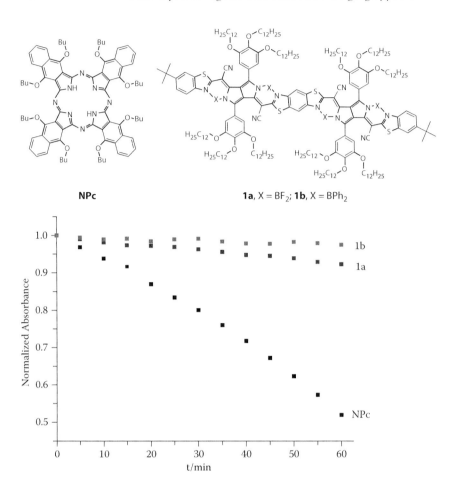

NPc   **1a**, X = BF$_2$; **1b**, X = BPh$_2$

Figure 5.3 Photo-bleaching test (halogen lamp, 68 W) of **1a** and **1b** against **NPc**. (Adapted with permission from G. M. Fischer et al., *Angew. Chem. Int. Ed.* **2011**, *50*, 1.)

Colorless, NIR-absorbing bis(pyrrolopyrrole) cyanine dyes **1a** and **1b** (or **64** in Chapter 3) are reported to be quite photostable [7]. The photostability of bis(pyrrolopyrrole) cyanine dyes were tested using a commercially available NIR dye (Sigma-Aldrich, CAS: 105528-25-4), namely 5,9,14,18,23,27,32,36-octabutoxy-2,3-naphthalocyanine (**NPc**) as a benchmark. Solutions of two respective dyes in toluene in 1 cm quartz cuvettes (3 mL) were illuminated with the emission of a halogen lamp for 1 h. Every 5 minutes, the absorption of the solutions was measured. Figure 5.3 displays the change of the absorption maxima after illumination of three samples. The absorbance of a solution of **NPc** decreased by 48%. In contrast, solutions of **1a** and **1b** exhibited an absorbance change of only 8% and 2%, respectively, after 60 minutes of illumination under identical

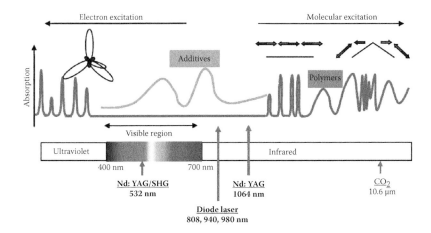

Figure 5.4 Light absorption of conventional polymers and laser wavelengths.

conditions. Thus, the colorless bis(pyrrolopyrrole) cyanine dyes are a promising new class of NIR absorbers, although more rigorous weathering or field tests need to be done.

The development of the laser as an industrial heat source has resulted in a range of applications that utilize the precise, controllable energy it delivers. For plastics processing, laser technology is more cost efficient than conventional processes. Nowadays, technologies for laser sintering, laser marking/printing, laser engraving, laser welding, and laser direct structuring have been well established in industry. Virtually any plastic can be laser processed, but material- and process-specific restrictions must be taken into account. Most of the commercial plastics do not absorb laser radiation in the region extending from the near-ultraviolet to the NIR. Conversion of laser energy into heat (of fusion) is therefore possible only if the polymer has been appropriately "laser sensitized" by addition of an additive. In absence of laser additives, therefore, most polymers can be processed only in far-ultraviolet light with excimer lasers, and in far-infrared light with CO_2 lasers (Figure 5.4). Diode lasers and Nd:YAG laser are relatively more available and can deliver a high power of light at the NIR wavelengths. Since most of conventional polymers do not absorb in the NIR region, it is necessary to use an additive or NIR absorber for laser processing of polymers.

Application profiles of laser welding typically include medical products, such as syringes, hose/tube connectors, pharmaceuticals packaging, pouches for liquids, in-line filters, catheter tips, balloon catheters and sensors, and electrical parts including miniature relay housings and dust- and watertight housings for electronic components. Evonik offers a range of laser-processable plastics and the corresponding technologies. In the

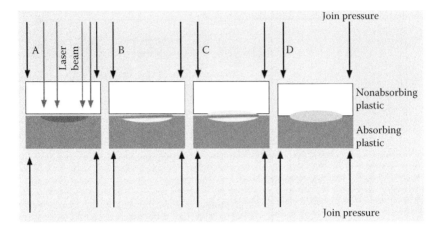

Figure 5.5 Stages of the laser welding process developed by Evonik.

Evonik laser welding process, the upper join partner is a laser-transparent thermoplastic, which absorbs little or no laser radiation and thus heats up very little (Figure 5.5). For a weld seam to be produced, the second join partner must absorb the laser radiation and thus can be dark colored by itself or doped with laser additives such as carbon black (approx. 0.3 wt%), metal oxides, or special dyes.

As shown in Figure 5.5, the laser passes through the upper join partner and is absorbed in the lower join partner (A). The laser energy is converted into heat, and a melt is formed (B). This now heats the upper join partner in the area of the seam to the extent that the material here also melts (C). Due to the external compaction pressure on the two join partners, the melt cannot escape, and the parts are welded (D). A drawback in this laser welding technology relates to the need for light-absorbing plastics, which must be either colored or doped.

A relatively new through-transmission laser welding technique for welding infrared transmissive and visually translucent plastics, called Clearweld™ developed by TWI and GENTEX in the United States [8], has been commercially available since January 2002. A wide range of commercial thermoplastics, such as polycarbonate, PMMA, polysulfone, PETG, LDPE, polyurethane, cyclic-olefin polymers, PEEK, clear polystyrene, transparent Noryl, polyester, nylons, and PVC, can be welded in virtually any joint geometry using the Clearweld process. The technology requires the use of colorless NIR dyes as a heat absorber, typically cyanine, squarylium, and the croconium dyes designed for use with lasers in the range of 900–1064 nm and specifically formulated for liquid dispensing and ink jet printing by Gentex Corporation. The absorber is preferably applied by spraying or dipping at the interface of the two substrates to be joined (Figure 5.6). The NIR dye at the interface between the two plastic

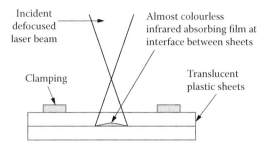

Figure 5.6 Schematic representation of the transmission laser welding Clearweld™ process using NIR dye.

parts acts as the site where the light from the laser is absorbed and converted into heat in a well-defined area. The area of heating, and hence joining, may be defined by either the size of the laser beam or the extent of the dye-containing region.

A variety of polymers in the form of sheets, parts, and films can be laser welded. The laser welding technology continues to develop and expands its applications in other areas, noticeably in fabric welding and tissue welding. The use of a laser to weld tissue in combination with a topical photosensitizing dye permits selective delivery of energy to the target tissue. A feasibility study using indocyanine green (IG) with the absorption peak at 780 nm was carried out with albino guinea pig skin [9]. IG was shown to bind to the outer 25 µm of guinea pig dermis and appeared to be bound to collagen. The optical transmittance of full-thickness guinea pig skin in the NIR was 40%, indicating that the Alexandrite laser (commonly used for hair removal) should provide adequate tissue penetration. Laser welding of skin in vivo was achieved at various concentrations of IG from 0.03 to 3 mg/cc using the laser at 780 nm, 250-µs pulse duration, 8 Hz, and a 4-mm spot size. IG dye-enhanced laser welding is possible in skin and with further optimization may have practical application.

The ideal NIR dyes for laser welding should have the following attributes: (1) a narrow absorption band near the laser wavelength with a high molar absorption coefficient, (2) very low or no absorption in the visible region of 400–700 nm, (3) good solubility in the host material, (4) good stability toward the processing method used, and (5) no formation of colored by-products. If it is intended for biomedical application, the nontoxicity property is a must. Considering all the known NIR dyes, the vast majority can be discounted on the grounds that they have pronounced visual color. Others can be disregarded because of their instability or their low absorption intensity. Therefore, it still remains a great challenge to date for one to develop ideal colorless NIR heat absorbers.

The applications of NIR-absorbing materials as discussed earlier have focused largely on exploiting the absorbance in the neutral state.

For those that may have multioxidation states, some applications only require the use of the materials in one particular single oxidation state such as Q-switching of lasers [10] and use as NIR-sensitive photocurrent detectors [11], rather than on the difference in absorbance as a result of a redox change.

5.2   Applications based on NIR chromogenic property

NIR chromogenic materials refer to those that can absorb in the NIR spectral region when being switched to one of bi- or multistationary states by external stimuli such as electrochemical reaction, light, heat, pressure, and so on. Accordingly, electrochromic, thermochromic, photochromic, piezochromic, and halochromic compounds are all potentially NIR chromogenic and may become technologically useful, in particular when they become highly absorbing at the telecommunication wavelengths (e.g., 1310 and 1550 nm). If so, these NIR chromogenic materials can in principle be used in devices for NIR attenuation and absorption or antireflection.

Variable optical attenuator (VOA) is an essential component in advanced wavelength division multiplexing telecommunication systems, which is used to adjust power variations caused by changes in source power, amplifier gain, and other components. Commercially available VOA devices operate mainly based on optomechanical and thermooptic (TO) systems and usually have response times of the order of milliseconds. VOA devices based on microelectromechanical system [12] and TO silica [13] and thermooptic polymers [14] are reported. In order to develop the planar VOA and other integrated photonic devices, various NIR electrochromic (EC) materials, especially metal-containing complexes, have been explored [15]. It has been known since the discovery of the Creutz–Taube ion that dinuclear complexes can show, in their mixed-valence state, an intervalence charge transfer (IVCT) transition from the electron-rich to the electron-poor metal terminus, which usually occur around the telecommunication bands of 1310 and 1550 nm and are very intense [16]. Thus, these dinuclear complexes are excellent candidates as NIR EC materials for VOA devices. This possibility was exploited by a number of research groups, in particular by Wang and coworkers, using **DCH-Ru** (Chapter 3) and polymers containing these active DCH-Ru units.

From the material point of view, the optical density should be as high as possible per active molecule and the reactive functional groups are needed in order to covalently incorporate active molecules into a polymer host. To this end, dinuclear ruthenium units (DCH-Ru) were incorporated into a single molecule, as represented by complexes **2a** and **2b** [17]. There are a number of ways to incorporate the metal complex moiety into

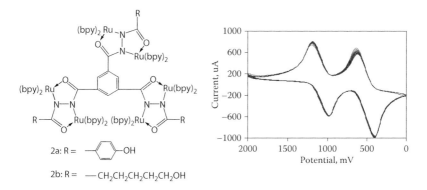

Figure 5.7 Dendritic complexes **2a** and cyclic voltammogram (20 cycles) of cross-linked film of complex **2a** on ITO in acetonitrile/0.1 M TBAH.

processable polymers, oligomers, and dendrimers. A common method is to first synthesize the corresponding ligand polymer and then introduce the metals through the exchange reaction with readily available metal complex [18]. The dendritic complexes **2a,b** were prepared by this ligand exchange method and were able to form thin films by casting or spin coating. To enhance the adhesion of the complex films on the electrode in the electrolyte medium, complexes **2a,b** were also incorporated in a polymer network by crosslinking with a triisocyanate compound. The crosslinking reaction for complexes **2a,b** was carried out under nitrogen at 120°C–140°C, and the completion of the crosslinking reaction could be monitored by IR for the characteristic band of isocyanate (2260 cm⁻¹). The cured complex films tightly adhered to the ITO electrode and were stable to a prolonged soaking in organic solvents such as *N,N*-dimethylformamide, acetone, acetonitrile, and chloroform. The electrochemical behaviors of the cross-linked film of complexes **2** (Figure 5.7) were almost identical to that of complexes **2**, indicating that the crosslinker did not disturb the electrochemical behaviors of active EC units.

From the device point view, there are at least two types of electrochromic VOA (ECVOA) devices. One operates in a reflective mode, such as device (a) in Figure 5.8, and another is in a transmission device (b and c, Figure 5.8). For device (a), it consists of top and bottom electrodes

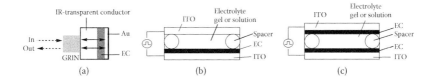

Figure 5.8 Reflection- and transmission-mode ECVOA devices.

sandwiching an inorganic or organic electrochromic material (e.g., WO_3 or PEDOT); one being gold or aluminum and the other being highly transparent at the wavelengths of interest, such as doped single crystal Si. To construct a fiber-optic device, a GRIN lens is used and bonded to the device. In the transmission device, either single EC layer (device b) or double EC layer (device c) can be employed. An electrolyte layer is used and could exist as a polymeric gel or a solution.

A dynamic attenuation of 15 dB with a switching response of 5 dB/s at 1500–1600 nm has been demonstrated in an ECVOA with a configuration of device (a) [15a]. Using a crosslinked thin film of **HO-OXA-Ru** (Figure 5.9) in the reflective ECVOA was also demonstrated [19]. The hydroxyl groups in the complex allow for the covalent bonding with a triisocyanate cross-linker (derived from trimethylolpropane and xylene diisocyanate), and the crosslinked film was NIR electrochromic with an absorption maximum at about 1550 nm (Figure 5.9). The reflective ECVOA is conceptually demonstrated using a simple setup shown in Figure 5.9, which consists of the EC film modified ITO electrode on glass strip that is coated with a reflective silver mirror. While ITO on glass is a suitable working electrode for visible applications, it begins to reflect the light from 1000 nm, weakening the detected signal strength. This is especially significant for detecting the optical signal at 1550 nm, as 35% of the light can be lost through reflection. However, this optical reflection could be harvested by working in

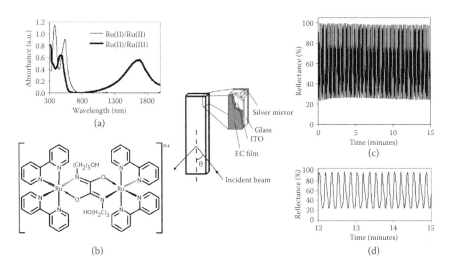

Figure 5.9 Left: (a) Absorbance spectra of the two oxidation states of **HO-OXA-Ru** and (b) its structure (PF_6^- counter anions not shown). Middle: A set up for a conceptual reflective ECVOA (θ = incident angle). Right: (a) Chronoamperometry of a silvered film cycling from –0.5V to 0.8V with a 5-s stepping time and (b) last 4-min expansion.

a reflective mode for optical attenuation, where the signal beam passes through the EC material and is reflected back by the ITO electrode.

Optical contrast between the two redox states was measured using a Harrick Scientific™ variable angle specular reflectance module and an in-house plexiglass cuvette holder. The coated ITO strip (8 × 40 mm) was submerged in a four-sided 1-cm quartz cuvette containing 0.1 M TBAH in acetonitrile as a supporting electrolyte solution. The MLCT absorption of the ruthenium complex in the polymer film (electrochemically reduced at −0.5V) was used as the background correction. Since the intensity of the reflected light depends on the angle of incident light, the signal strength was probed within a range of incident angles from 30° to 70°. The reflectance was measured with an incident angle of 56°, while the polymer film was electrochemically oxidized and reduced between the two oxidation states. It was found that the polymer films coated on bare ITO glass gave only 30% optical contrast or 1.6-dB attenuation at 1550 nm. Using the silvered ITO glass, an attenuation of 7 dB (or 80% optical contrast) at 1550 nm was readily achieved. Considering the thickness factor of active EC layer, optical attenuation for this EC polymer (with a thickness of 400 μm) in this reflective device is calculated to be 17.5 dB/μm at 1550 nm. In comparison with the transmissive device, the similar ruthenium complex gave the attenuation values of 0.05 dB/μm in a 100 μm (path length) solution-based device assembly [20] and 4.9–5.4 dB/μm in a solid film device [17c]. While some variation can be accounted for by the differences in molar extinction coefficients of the complexes used, it is clear that the device configuration also plays a role in performance. Chronoamperometric study showed excellent stability overextended switching cycles as well as retaining high attenuation with a 5-s stepping time and cycling potentials from −0.5 V to 0.8 V. Under these conditions, 70% optical contrast was achieved, which translates to a dynamic range of attenuation of 14.9 dB/μm of EC polymer film thickness.

A single-layer ECVOA with a configuration of device (b) based on thermally annealed WO_3 thin film in an electrolyte solution was disclosed in a patent [21]. The device was demonstrated with an attenuation range from 0.26 dB to 40 dB at 1310 nm and 1550 nm and could achieve an attenuation of 30 dB in 10 s. However, the reverse process, from the NIR "colored" to "colorless" state, took 10–15 min to recover about 80% of its transparency and up to 60 min for a full recovery. Compared to metal oxides, conducting polymers such as poly(3,4-ethylenedioxythiophene) (PEDOT) show a much faster switching response (ca. in the ms range) [22]. However, for a single-layer EC device, its absorption coefficient is still too low to allow for a VOA to operate in a transmission mode with a required dynamic attenuation over 20 dB at the 1550 nm wavelength. Increasing the thickness of EC layer can improve the attenuation range but significantly slow down the switching process or ion transport process.

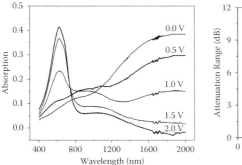
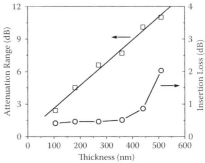

Figure 5.10 Vis–NIR absorption spectra of PEDOT film (140 nm thickness) on ITO in 0.1 M LiClO$_4$ propylene carbonate solution under different applied potentials (left). Effect of PEDOT film thickness on the attenuation range and optical loss at the bleached state of the solid–liquid ECVOA device under 2 V/0 V switching step (right). (Adapted with permission from J. D. Zhang et al., *Opt. Mater.* **2004**, *27*, 265.)

PEDOT as received is in the oxidized state and has a strong absorption in the NIR region (1000–2000 nm) with a peak maximum at about 1800 nm and an absorption coefficient of $\varepsilon_{ox} = 2.8 \times 10^4$ cm^{-1} at 1550 nm (Figure 5.10). When a negative potential is applied, PEDOT changes to its reduced state that has a weaker absorption in the NIR region. At the –2.0 V potential, PEDOT has nearly zero absorption (or $\varepsilon_{red} \sim 0$ cm^{-1} within the experimental error) at 1550 nm. Thus, a maximal power of optical attenuation in the transmission mode that PEDOT could offer relates to the difference between the absorption coefficients of the oxidized and reduced (or bleached) states, that is, $\Delta\varepsilon = \varepsilon_{ox} - \varepsilon_{red} = 2.8 \times 10^4$ cm^{-1}. Accordingly, the maximal attenuation per film thickness is calculated to be 27.7 dB/μm at 1550 nm.

It is expected that the thicker PEDOT film will expand the optical attenuation range at the expense of response time. A simple solution is to dope PEDOT film with an electrolyte such as LiClO$_4$. For practical application, the optical loss at the bleached state for the device should be as low as possible. As shown in Figure 5.10, optical attenuation increases proportionally to the film thickness. The optical loss of ECVOA device at the bleached state, as measured by its absorption when PEDOT is fully reduced, became significant when the film thickness was over 400 nm. For films under 400-nm thickness, optical loss was about 0.5 dB at 1550 nm. The optical loss mainly comes from the absorption of possibly residual oxidized PEDOT and the overtone absorption of the C-H bonds within PEDOT, PSS, and solvent molecules. When the thickness of the film is over 400 nm, the redox reaction becomes significantly slow and incomplete. As a result of incomplete reduction, the residual oxidized PEDOT is the main source of high optical loss. Accordingly, an optimized ECVOA device was fabricated, in which the PEDOT film was 440-nm thick and doped with 0.1 M LiClO$_4$, and showed a large attenuation range of 10.2 dB in over several hundreds of switching cycles, a moderate response time of 5 s for bleaching and 7 s for coloring and an optical loss of 0.86 dB [23].

ECVOA with a configuration of type (c) (Figure 5.8) works in the same principle as an electrochemical cell and employs two electrochromic materials that exhibit reversible and stable optical changes at the telecommunication wavelengths. When the device is switched between two different voltages, both electrochromic materials on each of the electrodes should be NIR absorbing under one applied voltage and become NIR transparent at another voltage. When the NIR light is directed through an ECVOA device, the light will either pass through or be absorbed at two distinct applied voltages and thus can be attenuated variably depending on the degree of the redox reaction or applied voltage. To demonstrate such a VOA device, the films on the working ITO electrode were prepared by crosslinking of complexes **2a** and **2b** with a triisocyanate compound [17c] and a thin layer of WO$_3$ was electrochemically deposited on the counter ITO electrode as a complementary material [24]. A polymer electrolyte gel was formulated with the composition of LiClO$_4$:PEO:propylene carbonate:ethylene carbonate in a ratio of 9:20:3:41 by weight. The gel electrolyte was cast onto the counter electrode and then assembled with the working electrode with a 0.1-mm spacer. The assembled devices were placed into a glove box for 3 days to allow complete gelation. The overall electrochemical process of such a device may be written as follows.

$$x\{[(Ru^{II}/Ru^{II}-L)_3]^{+6}(PF_6^-)_6\} + 3\,WO_3 + 3x\,LiClO_4 \rightleftharpoons x\{[(Ru^{II}/Ru^{II}-L)_3]^{+9}(PF_6^-)_6(ClO_4^-)_3\} + 3\,Li_xWO_3$$

transparent at 1550 nm   colored at 1550 nm   L = Ligand

The devices demonstrated rapid responses in the NIR region to the applied potential. For example, the device based on complex **2a** and WO$_3$ had an attenuation of 5.4 dB at 1550 nm with a switching time of 5 s (Figure 5.11).

A slightly different approach to anchoring NIR EC metal complexes onto the electrode was taken by Bignozzi and coworkers [25]. They introduced a carboxylate in NIR EC complexes such as **3**, which allowed the adhesion of these complex molecules onto transparent SnO$_2$:Sb or TiO$_2$ electrodes via the carboxylate anchor. Although the optical attenuation is expected

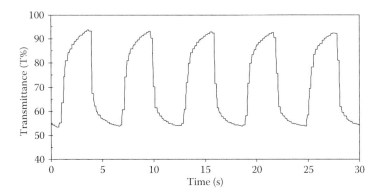

Figure 5.11 Variation of transmittance for ECVOA containing the films of cross-linked **2a** polymer and WO$_3$. (Adapted with permission from Y. Qi, Z. Y. Wang, *Macromolecules*, **2003**, *36*, 3146. Copyright (2003) American Chemical Society.)

to be low due to low content of active NIR materials adsorbed on the electrode, the switching times in these systems are relatively fast on the millisecond timescale.

Other NIR EC metal complexes with comparable or better properties are available for VOA applications. One obvious candidate comes from a series of Ni, Pd, and Pt dithiolene complexes **4**, which are well known to have intense NIR absorptions in the neutral and monoanionic states, but not in the fully reduced dianionic state [10,26]. Given the remarkable optical properties of these metal complexes, there should be great potential and plenty of opportunities by exploitation of dithiolene complexes for device applications.

The NIR electrochromic property of **DCH–Ru** complexes have also been explored for sensing the chemically and biologically important substances in aqueous media under physiological conditions [27]. It is known that tissue autofluorescence and light absorption by tissue and water are low between 900–1300 nm [28]. Since oxidation of **DCH–Ru** complexes to the mixed-valence state can also be done chemically with various oxidizing agents such as hydrogen peroxide, the presence of H$_2$O$_2$ in water could be detected by monitoring the NIR absorption of **DCH–Ru** complexes. Furthermore, H$_2$O$_2$ is an intermediate product from enzymatic glucose oxidation; therefore, glucose can also be detected indirectly and quantified under the physiological conditions (Figure 5.12). The sensing mechanism involves the oxidation of glucose by glucose oxydase (GOX) to produce H$_2$O$_2$, subsequent oxidation of a nuclear ruthenium complex to its Ru(II)/Ru(III) state, and detection of NIR absorbing signal. Complex **5** was designed to have a strong electron-donating NHPr group in order to

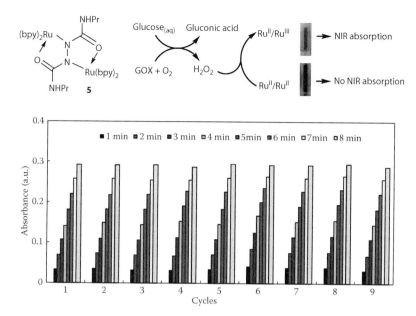

Figure 5.12 (See color insert.) Top: Proposed detection of glucose or hydrogen peroxide using dinuclear ruthenium complex. Cuvettes shown are ACN solutions of complex **5** in its respective oxidation states. Bottom: Normalized absorbance at 1150 nm of a film containing **5** on ITO at a given time after treatment with H_2O_2 (0.3 M) in TRIS solution. Film was reduced electrochemically (−0.5 V) in 9 cycles. (Adapted with permission from S. Xun et al., *Org. Lett.* **2006**, *8*, 1697. Copyright (2006) American Chemical Society.)

realize a blueshift in the MMCT absorption to 1150 nm, which falls in an ideal bio-window for sensing in an aqueous medium. Hydrogen peroxide readily oxidizes complex **5** to the Ru(II)/Ru(III) state, resulting in a significant increase in NIR absorption and color change from purple to yellow, although this visual effect is slow to appear to the naked eye. Complex **5** is sensitive enough to detect H_2O_2 at concentrations as low as 0.05 mM, which is much lower than the normal glucose levels in blood (typically in a range of 6–11 mM). To show a feasibility for sensor device application, complex **5** was doped in a crosslinked polyurethane film (0.58 μm) coated on ITO electrode glass, and the film-coated ITO plate was placed in a TRIS buffer solution (pH = 7.4) (2 mL), which is the same pH as human blood. A small amount (50 μL) of H_2O_2 was introduced to oxidize complex **5** doped in the film, while monitoring the increase in absorbance at 1150 nm simultaneously. After 8 min, a bias (−500 mV) was applied to the ITO plate after placing it in a TBAH/acetonitrile solution (0.1 M) to reduce the complex to its Ru(II)/Ru(II) state. These cycling experiments were repeated nine times with an excellent reproducibility (Figure 5.12). The crosslinked film tightly adhered to the ITO electrode and was stable to prolonged soaking in acetonitrile and TRIS buffer solution. There was no noticeable damage to the film on the ITO plate after many cycles of treatment of H_2O_2 and electrical reduction, demonstrating a potentially robust sensor device.

5.3 Photovoltaic applications using NIR organic materials

Photoactive materials are broadly defined as the materials capable of responding to light or electromagnetic radiation. Once a molecule has absorbed energy in the form of electromagnetic radiation, there are a number of routes by which it can return to the ground state, as illustrated by the Jablonski energy diagram, which typically shows radiative transitions involving the absorption and emission and nonradiative transitions involving several different mechanisms (Figure 5.13). Three nonradiative deactivation processes are significant here: internal conversion (IC), intersystem crossing (ISC), and vibrational relaxation. Examples of the first two can be seen in the diagram. Vibrational relaxation, the most common process for most molecules, occurs very quickly ($<1 \times 10^{-12}$ s) and is enhanced by physical contact of an excited molecule with other molecules

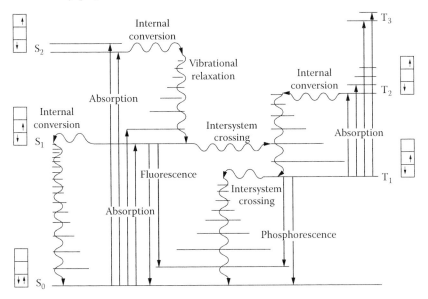

Figure 5.13 Jablonski energy diagram.

with which energy, in the form of vibrations and rotations, can be transferred through collisions. This photosensitization process means that most molecules in the excited state never emit any energy, rather transfer their energy to other molecules, before other deactivation processes can occur.

Using the NIR photosensitizing organic materials in dye-sensitized photovoltaic cells (DSCs) has recently received tremendous interests, aiming at increase of the power conversion efficiency (PCE) by capturing and converting more than 50% of solar energy that falls in the NIR spectral region (above 750 nm). DSCs are made from inexpensive materials, typically including a wide band-gap semiconductor anchored with a photosensitizing dye and a redox couple that are readily available in large quantity and easily processed at low cost [29]. DSCs work in a unique way in comparison with almost all other kinds of solar cells in that electron transport, light absorption, and hole transport are each handled by different materials in the cell [30]. When the dye absorbs light, the photoexcited electron rapidly transfers to the conduction band of the semiconductor, which carries the electron to one of the electrodes [31]. A redox couple, usually comprised of iodide/triiodide (I^-/I_3^-), then reduces the oxidized dye back to its neutral state and transports the positive charge to the platinized counter electrode [32].

CYC-B11   YD2-o-C8   Y123

Co(bpy)$_3$ Spiro-OMe TAD

Table 5.1 PCE, J_{sc}, V_{oc}, FF, and loss-in-potential for best-in class-DSCs

Dye	Couple	PCE (%)	J_{sc} (mA cm^{-2})	V_{oc} (mV)	FF (%)	Loss-in-potential (mV)
CYC-B11	I_3^-/I^-	11.5	20.1	743	0.77	850
YD2-co-C8	Co(bby)$_3$	12.3	17.7	935	0.74	775
Y123	**spiro-OMeTAD**	7.1	9.5	986	0.77	890

Several major advances in the design of dyes and electrolytes for dye-sensitized solar cells have led to a new record PCE of 12.3% [33]. The best-in-class DSCs are only based on a few best-performing materials, including the ruthenium complex **CYC-B11**, donor–π–acceptor dyes of **YD2-o-C8** and **Y123**, iodide redox couple, redox mediator Co(bby)$_3$, and hole conductor **spiro-OMeTAD**. The device characteristics are summarized in Table 5.1 [34]. The efficiency of a DSC can be improved in two ways, namely, extending the light-harvesting region into the NIR and lowering the redox potential of the electrolyte to increase open-circuit voltage (V_{oc}).

The slow recombination and relatively fast dye regeneration rates of the iodide redox couple have resulted in near-unity internal quantum efficiencies for a large number of photosensitizing dyes, providing the high external quantum efficiencies (Figure 5.14a). The study indicates that absorbing into the NIR region of the spectrum (e.g., 750–950 nm) increases the photocurrent density from 20 mA cm^{-2} to 30 mA cm^{-2} (Figure 5.14b). Calculation further shows that using a dye that absorbs around 950 nm, while still managing to generate and collect the charge carriers efficiently, could increase the current by over 40% (Figure 5.14c) [37].

However, finding one dye that absorbs strongly in the visible and NIR spectral region from 350 to 950 nm is rather difficult. Only a few NIR dyes with a maximal absorption over 700 nm have so far demonstrated good charge injection efficiencies in DSCs, although no NIR dye has yet independently achieved a V_{oc} greater than 460 mV in an electrolyte-based cell [38]. For example, a novel class of NIR-absorbing squarylium sensitizers with linearly extended π-conjugated structure has been developed for DSCs [38c]. The cells based on these dyes (e.g., **LSQa** with absorption maximum of 777 nm) exhibited a significant spectral response (IPCE%) in the NIR region over 750 nm in addition to the visible region (Figure 5.15). Nevertheless, application of deep-red and NIR dyes as a photosensitizer in DSCs becomes a fast-growing area of research and development [39]. Below are a few examples of deep-red dyes **6–12,** and their performance in DSCs are summarized in Table 5.2.

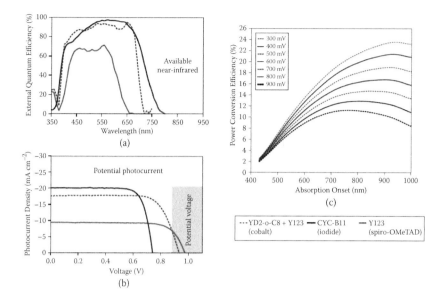

Figure 5.14 (a) External quantum efficiency versus wavelength. (b) Photocurrent density versus voltage for three DSCs containing different photosensitizing dyes and **CYC-B11**/iodide redox couple [35], the cosensitized donor–π–acceptor dye (**YD2-o-C8** and **Y123**)/cobalt redox couple [33], and a solid-state system comprised of the **Y123** dye and the hole conductor **spiro-OMeTAD** [36]. (c) Maximum obtainable power-conversion efficiencies versus absorption onset for various loss-in-potentials. (Adapted with permission from B. E. Hardin1 et al., *Nature Photon.* **2012**, *6*, 162.)

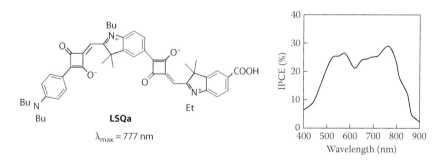

Figure 5.15 NIR squarylium dye-sensitized DSC exhibiting a spectral response above 750 nm. (Adapted with permission from T. Maeda et al., *Org. Lett.* **2011**, *13*, 5994. Copyright (2011) American Chemical Society.)

Table 5.2 Absorption and electrochemical properties of compounds **6–12** and performance of the dye-sensitized solar cells

	λ_{max} (abs) in solution (film) (nm)	E_g (eV)	Electrochemical HOMO (LUMO) (V versus NHE)	J_{sc}[a] (mA/cm^2) @P_{in}[b] 100 (mW/cm^2)	V_{oc}[c] (V)	FF[d] (%)	PCE (%)
6	680 (690)	1.77	1.20 (−0.57)	1.24	0.35	59	0.34
7	680 (640)	1.76	1.01 (−0.75)	4.54	0.38	58	1.38
8	636 (651)	1.92	0.98 (−0.78)	10.5	0.603	71	4.50
9[e]	660 (679)	1.54	0.43 (−1.11)[f]	8.6	0.591	73	3.70
9[g]	660 (679)	1.54	0.43 (−1.11)[f]	4.2	0.681	53	1.50
10	610 (632)	1.933	1.061 (−0.87)	11.76	0.464	67.4	3.70
11	597 (584)	1.881	1.149 (−0.73)	5.74	0.460	70.2	1.90
12	638 (687)	1.668	1.160 (−0.51)	2.49	0.317	66.2	0.5

[a] Short-circuit current.
[b] Incident radiation flux.
[c] Open-circuit voltage.
[d] Fill factor.
[e] Liquid cell.
[f] Data versus SCE.
[g] Solid cell.

6, λ_{max}^{abs} = 680 nm (1/1 MeCN/EtOH)
λ_{max}^{abs} = 690 nm (film)
λ_{max}^{pl} = 746 nm (1/1 MeCN/EtOH)

7, λ_{max}^{abs} = 680 nm (1/1 MeCN/EtOH)
λ_{max}^{abs} = 640 nm (film)
λ_{max}^{pl} = 742 nm (1/1MeCN/EtOH)

Aiming at broad absorption and high extinction coefficients, Jin and coworkers synthesized two NIR-absorbing dyes by covalently linking perylene imide with cyanine (**6**) and benzoindole (**7**) [40]. Both of them show two absorption bands with a low energy one at 680 nm. However, the maximal IPCE was only 3% for **6** and 11% for **7** and below 5% at the wavelength of 680 nm for both. Accordingly, the PCE of the DSCs based on **6** and **7** was rather low, 0.34% and 1.38%, respectively. The low efficiency may be due to the electron-withdrawing nature of the perylene imide group, which affects the polarity of the whole molecule, resulting in the unbeneficial transferring direction of photogenerated electrons and thus leading to a low overall photocurrent output.

8, λ_{vmax}^{abs} = 636 nm (EtOH)

λ_{vmax}^{abs} = 651 nm (film)

λ_{max}^{pl} = 659 nm (EtOH)

9, λ_{vmax}^{abs} ≈ 660 nm (EtOH)

λ_{max}^{abs} = 679 nm (film)

Another class of dyes with high extinction coefficients in the red and NIR spectral regions are squaraines and have been explored for use in DSCs [41]. Yum and coworkers designed squaraine dye **8** that absorbs the red with the peak maximum at 636 nm (ε = 158 500 dm³ mol⁻¹ cm⁻¹) and a portion of solar NIR radiation [42]. The DSCs fabricated with **8** had a maximal IPCE of 85% and PCE of 4.5% (J_{sc} = 10.50 mA/cm², V_{oc} = 603 mV, FF = 0.71). Using vertically oriented TiO_2 nanotube arrays, sensitized with regioregular **P3HT** plus the dye **8**, the solid state DSCs exhibited a broad spectral response with external quantum yields of up to 65% and PCE up to 3.8% [43]. Analog **9** in solution has a narrow absorption band in the red; however, its film shows a broadened and redshifted absorption band. The liquid cell based on **9** had a short-circuit current density of 8.6 mA cm⁻², an open-circuit voltage of 591 mV and a fill factor of 73%, giving rise to a relatively high PCE (3.7%). While solid-state DSCs sensitized with **9** had a short-circuit current density of 4.2 mA cm⁻², an open-circuit voltage of 681 mV, and a fill factor of 53%, leading to PCE of 1.5% [44].

10, λ_{max}^{abs} = 610 nm (MeCN)
λ_{max}^{abs} = 632 nm (film)

11, λ_{max}^{abs} = 597 nm (MeCN)
λ_{max}^{abs} = 584 nm (film)

12, λ_{max}^{abs} = 638 nm (MeCN)
λ_{max}^{abs} = 687 nm (film)

Three D–π–A type dyes **10**–**12** were synthesized and used as a sensitizer in DSCs. The device with **10** gave a higher IPCE between 400 nm to 800 nm than the other two dyes, and with a maximum value of 86% at 660 nm. The optimized device based on **10** achieved PCE of 3.7% (J_{sc} = 11.8 mA cm^{-2}, V_{oc} = 464 mV, *FF* = 0.674), being higher than those based on **11** (1.9%) and **12** (0.5%) [45]. The difference in PCE may be due to the more positive LUMO levels of compounds **11** and **12**, leading to the different orientation of dye molecules on the surface of TiO$_2$ and a different electron injection pathway.

With regard to the development of solely NIR dye-sensitized DSCs, another promising strategy for harvesting the whole spectrum is to use a combination of visible- and NIR-absorbing dyes or cosensitization. The current record-efficiency DSC employs a cosensitization strategy to boost absorption at a wavelength of 550 nm, as a result of an increased short-circuit current density rather than V_{OC} [33,46]. The most significant challenge of cosensitization using NIR dyes is to maintain a large V_{OC}, which requires that each dye adequately prevents recombination.

Since the first report of a molecular photovoltaic cell by Tang [47], the research and development of organic solar cells has been pushed by potential commercial uses as justified mainly by the low-cost fabrication process [48] and encouraged by the promising and steadily improving performance of bulk heterojunction (BHJ) solar cells [49]. Polymer BHJ solar cells based on regioregular poly(3-hexylthiophene) (**P3HT**, Figure 5.17) as an electron donor and [6,6]-phenyl-C$_{61}$-butyric acid methyl ester (PCBM) as an electron acceptor have a PCE typically in a range of 5–6%. However, **P3HT** has a band gap of ~1.90 eV and absorption wavelength below ~650 nm, which is a small portion of the whole solar spectrum (300–1500 nm, with maximal flux at 690 nm) [50]. To overcome this material limitation, the use of low band-gap polymers for replacement of **P3HT** in polymer BHJ solar cells is one solution and has been quite successful [51]. In comparison, low energy-gap small molecules or oligomers have several advantages over low band-gap polymers for use in solar cells [52].

In general, small molecule semiconductors are easy to reproducibly synthesize, functionalize, and purify. Moreover, crystalline compounds exhibit high charge mobility owing to their long-range order. With some

Table 5.3 Absorption and electrochemical properties of compounds **13–16** and performance of the BHJ solar cell devices

λ_{max} (abs) in solution (film) (nm)	Eg (eV)	Electrochemical HOMO (LUMO) (eV)	Donor/ PCBM Ratio (w/w)	$J_{sc}{}^a$ (mA/cm²) @$P_{in}{}^b$ 100 (mW/cm²)	$V_{oc}{}^c$ (V)	FF[d] (%)	PCE (%)
13 698 (~700)	1.41	−5.29 (−3.88)	1:4	1.95	0.75	34.4	0.75
14 616 (660, 742)	2.04	−5.03 (−3.0)	7:3	8.42	0.67	45	2.33
15 e (650, 720)	1.5	−5.2 (−3.7)	1:1[f]	9.2	0.75	44	3.00
15 e (650, 720)	1.5	−5.2 (−3.7)	10:10:2[g]	8.6	0.63	59	3.21
16 620 (660)	1.7	−5.2 (−3.4)	6:4[h]	10.0	0.92	48	4.40

[a] Short-circuit current.
[b] Incident radiation flux.
[c] Open-circuit voltage.
[d] Fill factor.
[e] No data supplied.
[f] **15**/PC$_{71}$BM.
[g] **P3HT**/PC$_{71}$BM/**15**.
[h] **16**/PC$_{71}$BM.

long alkyl chains and reasonably high molecular weights, they can be solution processed by casting, spin coating, and printing, suitable for use in large-area, low-cost solar cells. As the research on solar cells based on molecular sensitizers begins to emerge, the reports on the solar cells using low energy-gap small molecules are still scarce. The following are a few good examples (compounds **13–16**, Table 5.3).

13, λ_{max}^{abs} = 698 nm (THF)
λ_{max}^{abs} = 698 nm (film)
λ_{max}^{pl} = 700 nm

14, R = t-Boc
λ_{max}^{abs} = 616 nm (CHCl₃)
λ_{max}^{abs} = 660, 742 nm (film)

15: R = CH₂CH(Et)Bu
λ_{max}^{abs} = 650, 720 nm (film)

16, R = CH₂CH(Et)Bu
λ_{max}^{abs} = 620 nm (solution)
λ_{max}^{abs} = 660 nm (film)

Sun and coworkers reported a novel NIR dye (**13**) containing the D–π–A–π–D structure that have the absorption onset at 880 nm or optical energy gap of 1.41 eV, the HOMO and LUMO of −5.29 eV and −3.88 eV, respectively, as calculated from the redox potentials. The best device performance was achieved with a blend of **13** and PCBM in a ratio of 1:4 by weight, showing a PCE of 0.75% with an open-circuit voltage (V_{oc}) of 0.75 V and a short-circuit current density (J_{sc}) of 1.95 mA/cm^2 under AM 1.5 (Air Mass 1.5) solar simulation (100 mW/cm^2) [53].

The NIR-absorbing diketopyrrolopyrrole-based dye **14** was used as an electron donor in molecular BHJ solar cells [54]. The absorption of its film is broad, with two bands centered at 660 and 742 nm, due to the solid-state aggregation (Figure 5.16). The solar cell device using **14**/PC$_{61}$BM (ratio 70:30) gave a PCE of 2.33% with V_{oc} of 0.67 V, J_{sc} of 8.42 mA/cm^2, and

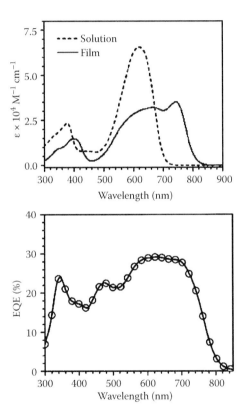

Figure 5.16 Absorption spectra of compound **14** in chloroform and as film on a quartz substrate. External quantum efficiency (EQE) curve for BHJ photovoltaic device using 70:30 blend of **14**:PCBM. (Adapted with permission from A. B. Tamayo et al., *J. Phys. Chem. C*, **2008**, *112*, 11545. Copyright (2008) American Chemical Society.)

fill factor of 0.45. This device exhibits external quantum efficiencies close to 25% at 343 nm and close to 30% between 550 and 750 nm (Figure 5.16), the former being mainly due to the PCBM acceptor, while the latter is attributed to the donor material **14**. The analog **15** is more thermally stable than **14** and also performs better, as evident by a solar cell with a PCE of 3.0% [55]. This better performance is attributed to the high degree ordering of **15** in the blend films and the balanced hole and electron mobility of 1.0×10^{-4} and 4.8×10^{-4} cm^2/Vs, respectively. The same compound has also been doped in **P3HT** to extend the absorption breadth, because its energy levels lie between those of **P3HT** and $PC_{71}BM$. The devices based on the annealed blend of **P3HT**:PC_{71}**BM**:**15** (10:10:2 by weight) showed a PCE of 3.21%, virtually the same as the molecular BHJ solar cell made of **15** and $PC_{71}BM$ [56]. Considering the correlation between the HOMO level of the donor and open-circuit voltage of the device [57], further tuning of the energy gap led to the design and synthesis of an NIR-absorbing dye **16** [58]. The benzofuran substituent in **16** sets the HOMO level at 5.2 eV, which results in a high V_{oc} of 0.92 V and the highest PCE of 4.4% reported by then for molecular BHJ solar cells.

In the field of research on polymer solar cells, there were two major breakthroughs in the 1990s. One of the major breakthroughs was made by independent observation of an extremely fast photoinduced electron transfer process of around 50–100 fs between a conjugated polymer and fullerene derivatives such as PCBM by Heeger et al. and Yoshino et al. [59]. Another was the realization of the first efficient bulk heterojunction polymer solar cells in 1995 by the groups of Heeger and Friend, using a blend of donor polymer and fullerene derivative [60]. Polymer–fullerene systems currently dominate the field of high-efficiency polymer solar cells. In 2005, researchers further discovered that the morphology, in particular the donor–acceptor phase separation, also plays a critical role in achieving proper charge transport channels for collecting the electrons and holes [61].

The development of polymer solar cells has always been accompanied by innovations in materials science. Early work by Yu et al. on the first polymer solar cell using the blend of **MEH-PPV** (Figure 5.17) and C_{60} and its derivatives has opened up a new era of polymer materials for use in solar energy conversion [60a]. After significant optimization, researchers achieved PCEs of more than 3.0% for PPV-based solar cells [62]. However, further improvement was limited by the relatively low hole mobility and narrow light absorption range. Soluble polythiophenes, such as **P3HT** [63], with their higher hole mobility [64] and, therefore, a broader spectrum coverage than **MEH-PPV**, have become a commonly used donor material in polymer solar cells in the 2000s. Morphology optimization has provided PCEs of 4%–5%, thus attracting worldwide interests in polymer solar cells [61].

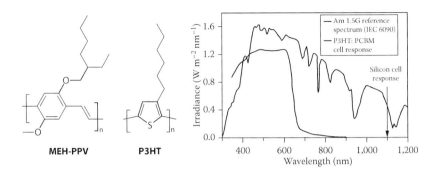

Figure 5.17 Chemical structures of two commonly used donor polymers, **MEH-PPV** and **P3HT**. Comparison between solar spectrum and the photoresponse of a **P3HT**:PCBM solar cell.

The efficiency of a solar cell is given by $\eta = V_{oc} \times J_{sc} \times FF$, where V_{oc} is the open-circuit voltage, J_{sc} is the short-circuit current density, and FF is the fill factor. Knowledge of the link between the design of a polymer and these parameters has been significantly enhanced over the past decade. Materials innovation is one of the major forces currently driving the performance of polymer solar cells. The key issues of polymer design include [65] engineering the band-gap and energy levels to achieve high J_{sc} and V_{oc}, enhancing planarity to attain high carrier mobility, and materials processability and stability [65]. All of these issues are correlated with each other.

A common strategy for achieving high J_{sc} is to narrow the band gap (<1.8 eV) for a broader coverage of the solar spectrum [65,66]. Taking **P3HT** as an example, there is a large mismatch between the solar spectrum and the EQE spectrum of the representative BHJ solar cell using **P3HT** and PCBM, since **P3HT** has a band gap of 1.9 eV and absorbs the light only up to 700 nm (Figure 5.17). Thus, in order to further broaden the absorption of donor polymers, polymers are usually designed by introducing an alternating donor–acceptor structure, stabilizing the quinoid structure, controlling the polymer chain planarity, and tuning the effective conjugation length. The alternating donor–acceptor structure facilitates electron delocalization and the formation of low band-gap quinoid mesomeric structures over the polymer backbone [65a,67].

One of the most successful examples is **PCPDTBT**, whose absorption extends up to 900 nm (1.4 eV). Solar cells made from this polymer have shown an initial efficiency of around 3% [68]. By incorporating alkane-dithiol additives, researchers were able to achieve efficiencies of around 5.5% [69]. Leclerc et al. reported a low band-gap polymer **PCbzTBT** and a PCE of 3.6% based on this polymer [70]. Further increase of PCE to 6.1% was accomplished by incorporating a titanium oxide layer as an optical spacer in the device [71]. A variation of this donor–acceptor structure, comprising an acceptor-based side chain and a donor-based main chain, was also proven to be valid in making low band-gap polymers [72].

The quinoidal form is energetically less stable because of its smaller band gap. Therefore, making the quinoidal form more stable would realize the band gap narrowing in a conjugated system. The most successful examples are those reported by Yu et al., which are composed of thieno[3,4-*b*]thiophene (TT) that can stabilize the quinoidal structure through a fused thiophene ring and benzodithiophene (BDT) alternating units and have band gaps of around 1.6 eV [51b,66,73]. The polymer **PTTBDT** represented the first donor in BHJ system capable of reaching PCEs of 7%–8%. Following this work, PCEs of more than 7% were frequently reported with either new materials or novel device optimization techniques [74].

It must be pointed out that, in addition to narrowing the band gap, other parameters, such as carrier mobility, intermolecular interaction, and molecular chain packing, also affect J_{sc}. For example, **PDTSBT** is an analog of **PCPDTBT** made by replacing the bridging carbon atom with a silicon atom [15b,75]. Due to its higher crystallinity than **PCPDTBT**, **PDTSBT** has improved hole mobility, leading to a higher value of J_{sc}.

Regarding the fill factor, it appears not directly related to the structural property of polymers. However, Fréchet et al. showed that the side-chain patterns for *N*-alkyl thieno[3,4-*c*]pyrrole-4,6-dione (TPD)-based polymers can affect the fill factor [76]. It is reasoned that side-chain tuning affects π-stacking of polymer chains, polymer crystallinity, and material miscibility, which in turn affects the *FF*. A solar cell based on **PBDTTPD** was reported to have a PCE of 7% [74e].

For solar cell applications, narrowing the band gap or increasing the NIR absorption of donor polymers alone does not always lead to enhancement of the PCE. Poly(5,7-bis(4-decanyl-2-thienyl)thieno[3,4-*b*]diathiazole-thiophene-2,5) (**PTTT**, Figure 5.18) was reported to have an E_g of 1.01 eV and absorb the light from 330–1220 nm as thin film. The photovoltaic devices based on the **PTTT**/PCBM blend show the photosensitivity response up to 1100 nm, J_{sc} of 0.83 mA/cm^2, V_{oc} of 0.35 V, and *FF* of 38.6%, but rather low PCE (0.07%–0.11%) [77]. However, the same polymer has been successfully utilized in a visible–NIR photodetector device with the performance parameters that are comparable to or even better than photodetectors fabricated from inorganic materials [78].

The development of optoelectronic devices for sensing IR radiation is critical for a diverse variety of applications such as image sensing, optical communications, environmental monitoring, remote control, day- and night-time surveillance, chemical/biological sensing, and spectroscopic and medical instruments [79]. Most widely used NIR detectors depend on electron–hole generation in low band-gap semiconductor structures

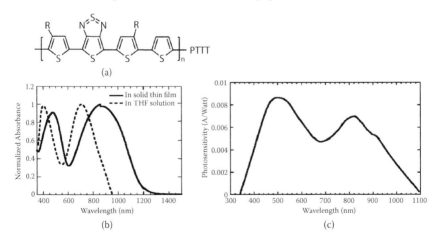

Figure 5.18 (a) Chemical structure of **PTTT**. (b) Normalized absorption spectra of **PTTT** in solid thin film and in the THF solution (5 x 10⁻⁵). (c) photosensitivity of a photovoltaic device based on the **PTTT**/PCBM (1:3) blend. (Adapted with permission from Y. Xia et al., *Appl. Phys. Lett.* **2006**, *89*, 081106.)

by incident radiation. These detectors exhibit good signal-to-noise performance and very fast response. However, in order to achieve this, the semiconductor components of the detector need cryogenic or thermoelectric cooling. Thermal generation of carriers in a low band-gap semiconductor, impurity- and defect-mediated recombination, or thermally activated intersubband transitions (i.e., in quantum well detectors) are limitations of conventional semiconductor photon detectors. In general, GaN-, Si-, and InGaAs-based detectors are used for the three important subbands, 0.25 μm to 0.4 μm (UV), 0.45 μm to 0.8 μm (visible), and 0.9 μm to 1.7 μm (NIR), respectively. The detectivities of silicon photodetectors are ~4 × 10^{12} Jones (1 Jones = 1 cmHz$^{1/2}$/W). The typical detectivities of InGaAs photodetectors are greater than 10^{12} Jones when cooled to 4.2 K. At present, individual sensors are required to perform the so-called two-color sensing at the different wavelengths within the UV to NIR spectral region. Thus, it is highly desirable to use one material in a photodetector that can operate over the broad spectral range from the UV to the NIR with high detectivity and without a need for cooling.

In 2009, Xiong, et al. reported the **PTTT**-based photodetectors that exhibited spectral response from 300 nm to 1450 nm, with detectivity greater than 10^{13} Jones at wavelengths from 300 nm to 1150 nm and greater than 10^{12} Jones from 1150 nm to 1450 nm and a linear dynamic range over 100 decibels at room temperature [78]. When being compared with Si and InGaAs photodetectors, the polymer photodetector performs similarly or even better than Si and InGaAs photodetectors in terms of photodetectivity and spectral response range (Figure 5.19). The detectivity-wavelength profile for **PTTT**-based photodetector was obtained from the measured photoresponsivity data with absolute magnitude determined by points A and B. The optimized photodetector has a device configuration of ITO/PEDOT/PS-TPD-PFCB/**PTTT**:PC$_{60}$BM/C$_{60}$/Al and uses a crosslinkable TPD polymer and C$_{60}$ to form the blocking layers. High photodetectivity is mainly attributed to the very low dark current (and noise), which is due to low-lying HOMO level of low band-gap polymer.

Another low band-gap polymer (**PTT**, E$_g$ = 1.3 eV, Figure 5.20) was tested for use in a NIR photodetector [80]. The polymer absorbs from 600 nm to 900 nm. The photodetector exhibited the EQE over the same spectral region with the highest EQE of ~ 40% at 850 nm at −5 V bias (Figure 5.20). The device's figure of merit (NEP), or the minimum impinging optical power that a detector can distinguish from the noise, was 3.85×10^{-12} W/Hz$^{1/2}$ at 850 nm at 4 kHz (at 0 V), being still inferior to those of commercial NIR photodetectors.

A polymer, **PDAPBDT** (Figure 5.21), containing 2,5-diazapentalene (DAP) and BDT in the repeat unit has a narrow band gap of 1.19 eV. The solar cell based on **PDAPBDT** performs poorly in a BHJ solar cell with a rather low power conversion efficiency, but is promising for use in NIR

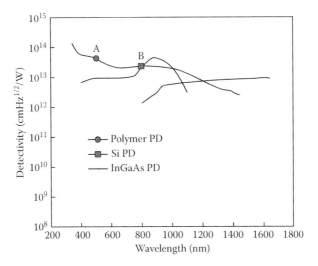

Figure 5.19 Detectivities of Si-, InGaAs-, and **PTTT**-based photodetectors. (Adapted with permission from X. Gong et al., *Science*, **2009**, *325*, 1665.)

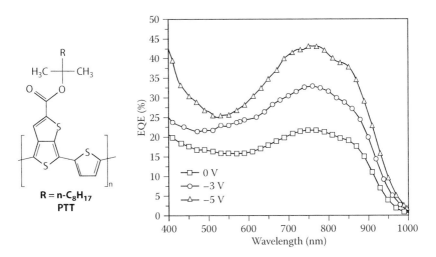

Figure 5.20 Low band-gap polymer **PTT** and the EQE of **PTT**-based photodetector versus wavelength at various biases. (Adapted with permission from Y. Yao et al., *Adv. Mater.* **2007**, *19*, 3979.)

photodetectors [81]. The device configuration is ITO/PEDOT:PSS (35 nm)/ active layer (100 nm)/Al (100 nm), and the active layer comprises a blend of **PDAPBDT** and $PC_{61}BM$ in a weight ratio of 1:3. The dark current density was 3.27×10^{-9} (A/cm^2) at bias of -0.1 V and even lower than that of the reported high-detectivity photodetector based on a narrow band-gap

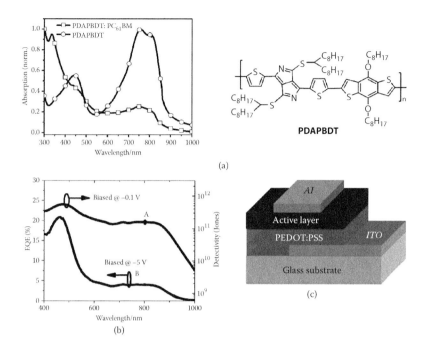

Figure 5.21 (a) Chemical structure of **PDAPBDT** and absorption spectra of films of **PDAPBDT** and **PDAPBDT** with PC$_{61}$BM (1:3 weight ratio). (b) EQE and detectivity at the wavelength of 400–1000 nm for **PDAPBDT**-based photodetector. (c) Schematic layout of the ITO/PEDOT:PSS/active layer/Al photodetector on a glass substrate. (Adapted with permission from G. Qian et al., *Chem. Mater.* **2012**, *24*, 2364.. Copyright (2012) American Chemical Society.)

polymer [79], which may be attributed to, at least in part, the lower HOMO level of **PDAPBDT.** The photocurrent spectral response covered a wide range from 400 nm up to 1000 nm, matching well with the absorption profile of the film of **PDAPBDT** and PC$_{61}$BM mixture (Figure 5.21). The EQE and detectivity were calculated at 800 nm (points A and B in Figure 5.21) as 4.0% and 1.6 × 10^{11} Jones, respectively. The relatively low EQE value may be due to the large domain size of the active layer (about 100 nm) and the close LUMO levels of **PDAPBDT** and PC$_{61}$BM that could lead to insufficient exciton dissociation. At the wavelengths from 400 nm to 850 nm, the device exhibited a photodetectivity greater than 10^{11} Jones, being comparable to previously reported polymer photodetectors [80,82].

An NIR-sensitive heterojunction *n*-TiO$_2$/Dye/*p*-CuSCN, where Dye denotes a NIR absorbing dye, was examined for a possibility of using dye-sensitization for NIR radiation detection. Several dyes were selected and tested in the dye-sensitized cells, but the responsivity was rather lower and response was slow compared to silicon photodetectors (Table 5.4) [83].

Table 5.4 Responsivities (*R*) of the
dye-sensitized photodetectors at
peak absorption wavelengths

Dye	λ_{max} (nm)	R (mA/W)
IR820	866	0.3
BR-IR820	876	1.1
IR783	808	0.4
MC-IR792	812	2.7
IR820-IR1040	858	1.0
BR-IR1040	1056	0.3

Perhaps, using highly efficient dyes, such as **CYC-B11, YD2-o-C8,** and **Y123** (Table 5.1), one would obtain a dye-sensitized NIR photodetector with better performance at a specific wavelength (e.g., 800 nm).

Photoresistors are one type of photodetectors that utilize the photoconductivity of a material. Due to the absorption of electromagnetic radiation such as visible light, ultraviolet light, infrared light, or gamma radiation, a material could become more electrically conductive, which is an optical and electrical phenomenon commonly called *photoconductivity* [84]. When light is absorbed by a semiconductor material, the number of free electrons and electron holes changes and raises its electrical conductivity. To cause excitation, the light that strikes the semiconductor must have enough energy to raise electrons across the band gap, or to excite the impurities within the band gap. When a bias voltage and a load resistor are used in series with the semiconductor, a voltage drop across the load resistors can be measured when the change in electrical conductivity of the material varies the current flowing through the circuit. When a photoconductive material is connected as part of a circuit, it functions as a resistor whose resistance depends on the light intensity. Some photodetector applications in which photoresistors are often used include camera light meters, street lights, clock radios, and IR detectors.

Classic examples of photoconductive materials include (i) poly(vinyl carbazole), squaraine and azo dyes, and vanadyl phthalocyanine [85], used extensively in photocopying (xerography); (ii) lead sulfide, used in IR detection applications, such as the U.S. Sidewinder and Russian Atoll heat-seeking missiles; and (iii) selenium, employed in early television and xerography. One typical example is the use of NIR dyes coated on the surface of the silver halide microcrystals that are only sensitive to ultraviolet and blue light, in order to make photographic materials sensitive to IR light by injecting electrons into the conduction band of the silver halide when the dyes absorb NIR light [86].

Thin-film devices with a configuration of Al/single-wall carbon nanotube (SWNT)-polymer/indium tin oxide were fabricated and exhibited the promising photoconductive response in a broad spectral range, typically from 300 to 1600 nm [87]. SWNTs containing both semiconducting and metallic carbon nanotubes are finely dispersed in a polymer film, either **MEH-PPV** or poly(3-ocylthiophene-2,5-diyl). These devices utilize a number of unique properties of SWNTs, including the intrinsic NIR light harvesting property, the electronic transport properties in combination with that of the polymer matrices, and probably the charge/energy transfer processes between the nanotubes and the polymers. In the NIR range, by photoexciting selectively semiconducting SWNTs (transition $v^s_1 \rightarrow c\ ^s_1$), bounded electron ($e$) and hole ($h$) pairs known as *excitons* are formed [88]. They eventually dissociate into free e and h under electric field, either externally applied or built-in due to the difference in the work function of Al and ITO. This mechanism is consistent with a related photoconductivity experiment on individual semi-SWNTs in a field-effect transistor configuration [89].

The unusual NIR-photoconductive properties of metal dithiolene complexes have been investigated with the aim of developing wavelength selective air-stable photodetectors. These metal (Ni, Pd, Pt) dithiolene complexes show intense absorptions at about 1000 nm with high molar extinction coefficients and are quite thermally and photochemically stable. Although the preliminary study revealed only a weak photoconductivity for a Ni–dithiolene complex at room temperature [90], it is still desirable to investigate other metal dithiolene complexes for potential use in photodetectors. For example, one of NIR Pt–dithiolene complexes was examined for photodetector application [91], which exhibited an interestingly high efficiency at about 1000 nm only (Figure 5.22). Therefore, this

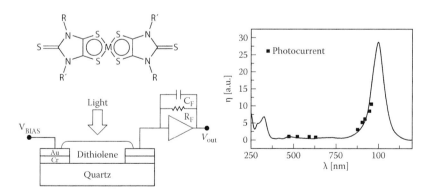

Figure 5.22 A planar photodetector prototype containing a dithiolene (M = Pt; R = Et, R′ = pentyl). Efficiency (η) from photocurrent measurements (dots) as a function of the incident wavelength. The UV–Vis–NIR spectrum recorded in chloroform solution is superimposed. (Adapted with permission from M. C. Aragoni et al., *Inorg. Chem. Commun.* **2002**, *5*, 869.)

device acts as a wavelength-selective NIR photodetector that is almost blind to visible light.

5.4   Transistor applications using low band-gap polymers

Ambipolar organic thin-film transistor (OTFT) is an important electronic device, in which organic low band-gap materials may find a niche application. Ambipolar OTFT with high hole and electron mobilities have attracted considerable attention due to their application in complementary metal-oxide semiconductor (CMOS) digital integrated circuits [92]. Considering the band-gap requirement for a suitable transistor material, the low-lying LUMO and high-lying HOMO levels would facilitate the efficient injection and transport of electrons and holes and, therefore, an ultra-low band-gap organic material is desirable and incorporation of both strong donor and strong acceptor in molecular or polymer design is a viable route to realize effective ambipolar field effect transistors.

To date, most of known conjugated compounds and polymers exhibit p-type mobility. For example, **PDPPTTT** (Figure 5.23) predominantly shows p-type mobility up to 0.94 cm^2 V^{-1} s^{-1} in organic thin-film transistors

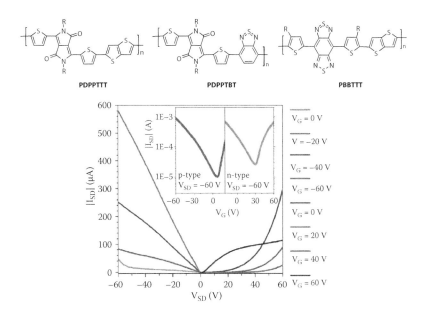

Figure 5.23 (See color insert.) p-Type **PDDPTTT** and ambipolar **PDDPTBT** and **PBBTTT** polymers. Transfer (inset) and output characteristics of a **PBBTTT** transistor. (Adapted with permission from J. Fan et al., *Adv. Mater.* **2012**, *24*, 2186.)

[93], but when the thieno[3,2-*b*]thiophene (TT) donor unit is replaced with a benzothiadiazole (BT) acceptor unit, the resulting polymer **PDPPTBT** becomes a typical ambipolar semiconductor [94]. **PDDPTBT** has LUMO of −4.0 eV and the HOMO of −5.2 eV, which are reasonably favored energy levels for stable hole and electron transport. Accordingly, the ambipolar TFT based on **PDDPTBT** exhibited an electron mobility of 0.40 cm^2 V^{-1} s^{-1} and a hole mobility of 0.35 cm^2 V^{-1} s^{-1}. Replacing the DPP unit with an even stronger acceptor of benzobisthiadiazole (BBT) is expected to further narrow the band gap of the resulting polymer such as **PBBTTT**, which should enable easier injection for holes and electrons. **PBBTTT** absorbs in the NIR region with the absorption onset of around 2200 nm in the solid state, which corresponds to an optical band gap of 0.56 eV [95]. For the electrochemical data, the LUMO energy level of **PBBTTT** was calculated to be −3.80 eV, whereas the HOMO value of −4.36 eV was determined from the difference between the LUMO and the optical band gap. This low band-gap polymer in a transistor device exhibits nearly balanced hole and electron mobilities as high as 1.0 and 0.7 cm^2 V^{-1} s^{-1}, respectively. The transfer and output characteristics of a typical transistor, annealed at 240°C, are shown in Figure 5.23 with transfer curves in the inset. By plotting $I_{DS}^{1/2}$ and I_{DS} as a function of V_{GS}, the linear slope of $I_{DS}^{1/2}$ versus V_{GS} is obtained for a large range of gate voltages, particularly at low gate voltages where saturation is observed for p-type transport, and for all gate voltages for n-type transport. High-performance ambipolar behavior can be clearly observed in both the transfer and output characteristics, with currents in the hundreds of microamperes. Using this ambipolar polymer, inverters were demonstrated with a gain of 35, which is much higher than values usually obtained for unipolar logic. **PBBTTT** probably represents a conjugated polymer with the smallest band gap ever reported to date. Besides the application in ambipolar transistors, this unique low band-gap polymer is expected to be a promising material for use in NIR photodetectors.

5.5 Applications based on NIR fluorescent property

The NIR fluorescence property of organic materials can be utilized in many ways for biomedical, sensor and telecommunication, and other emerging photonic applications. For example, NIR fluorescent dyes have been used to label biomolecules because of the low interference from solvents and the biological matrix in this region [96,97]. The light emitting diodes (LEDs) using NIR emissive organic materials are of particular interest because of a number of potential applications in areas such as IR signaling and displays, telecommunications, and bio-sensing [98–100]. In comparison with NIR-absorbing organic materials, NIR fluorescent organic materials are less available and even scarce for useful application in LEDs. In the past decade, organic materials that are used in OLEDs showing detectable NIR electroluminescence (EL) mainly include organic ionic dyes, rare-earth (lanthanide) complexes [101], transition-metal complexes [102], and low band-gap small molecules or polymers [103]. Although good progress has been made in all-organic LEDs with emissions below 1000 nm, it still remains scientifically and technically challenging to realize highly efficient all-organic LEDs that emit at the wavelengths beyond 1000 nm without any visible light. From the material side, one needs to further reduce the energy gap of the emitters, such as the use of a stronger electron acceptor of benzobisthiadiazole in the D–π–A–π–D type of emitters to redshift the emission wavelength. On the device side, one needs to use a multilayer structure to enhance the EL property and block off any visible emission.

H₂Pc

Q band: λ^{abs}_{max} = 630 nm

λ^{pl}_{ma} > 800 nm

DCM

λ^{abs}_{max} = 470 nm

λ^{abs}_{max} = 630 nm

In 1996, Fujii et al. firstly reported the NIR EL of metal-free phthalocyanine (**H₂Pc**) [104], in a device with the configuration of ITO/**H₂Pc**/Mg:In. The devices showed two distinct emission peaks at around 480 nm and above 800 nm, contributed to the transition of the B band and Q band, respectively. Later, the same group investigated the metal cathode-dependent EL characteristics of the devices based on blend film of **H₂Pc** and 4-(dicyanomethylene)-2-methyl-6-(*p*-dimethylaminostyryl)-4H-pyran (**DCM**) [105]. Due to the different work functions of metal cathodes of Al (−4.2 eV) and Mg (−3.7 eV), the device with Al cathode emitted mainly at 900 nm at very low temperature, presumably with very low quantum efficiency. Because the absorption spectrum of **DCM** is overlapped with the emission spectrum of the B band of **H₂Pc** and the emission spectrum of **DCM** is overlapped with the absorption spectrum of the Q band of **H₂Pc**, an efficient energy transfer from the B-band emission to **DCM** and then to the Q band of **H₂Pc** may occur in the blend film of **H₂Pc** and **DCM**. Consequently, nearly exclusive NIR EL was obtained from the devices containing the blend film, due to effective quenching of the B-band emission.

LDS821

λ_{max}^{el} = 800 nm

TCC

λ_{max}^{abs} = 722 nm

λ_{max}^{pl} = 735 nm

λ_{max}^{el} = 748 nm

ADS830AT

λ_{max}^{abs} = 812 nm

λ_{max}^{pl} = 830 nm

Some ionic dyes show relatively good NIR fluorescent properties, in terms of the emission wavelength and efficiency, and have been explored for use in NIR LEDs. Suzuki reported NIR EL from the devices based on ionic dye **LDS821** that was doped in poly(*N*-vinylcarbazole) (PVK) and 2-(4-biphenylyl)-5-(4-*t*-butylphenyl)-1,3,4-oxidiazole (PBD) [106]. Because the energy level of **LDS821** is between that of PVK, this dye can act as an effective carrier-trapping and radiative-recombination center, resulting in almost exclusive NIR emission at 800 nm from the LED. The device efficiency was very low (0.015%) when the concentration of PBD was 30% by weight. However, by applying reverse bias repeatedly over time, the EL quantum efficiency could reach 1%, due to the enhanced electron and hole injection caused by the alignment of the ionic dye molecules along the bias field. The EL properties of IR1051 were evaluated by doping the ionic dye (1 wt%) and PBD (30 wt%) in PVK [107]. The EL spectrum showed two emission peaks at 990 and 1100 nm, which were assigned to the aggregated dyes and monomeric dye, respectively. Besides the NIR emission, the device also emitted strongly the visible light that was accounted for more than 60% of the total emission intensity. Similar ionic dyes (e.g., IR1050, IR1051, IR1061, and S0447) were evaluated for use in NIR OLEDs [108]. Generally speaking, the devices made of these ionic dyes emit rather weak NIR light at around 1100 nm with very low quantum efficiency (e.g., 0.036%) and as well much stronger visible light. Some cyanine dyes tend to form NIR fluorescent J-aggregates, such as the aggregates of **TCC** dye in polyimide film [109]. The absorption of **TCC** J-aggregate in polyimide film appeared at 767 nm, and a narrow EL band at 815 nm was observed with a low quantum efficiency of 0.01%. The EL spectrum of cyanine dye **ADS830AT** doped in PVK displayed just one emission peak at 890 nm [110]. To improve the emission efficiency of the NIR LEDs by balancing the charge transport and fulfilling the energy cascade transfer, CdSe/CdS quantum dots were doped in the emissive layer and the LED device that produced with the highest quantum efficiency of 0.02% reported for the dye-based NIR OLEDs [111]. The authors claimed that CdSe/CdS quantum dots act as a good electron transport material that can balance the electron and hole transport in the emissive layer, thus leading to the improvement of the device emission efficiency.

The π-conjugated donor–acceptor compounds are good candidates for use in LEDs as emissive materials, since their energy gap can be easily tuned and they exhibit relatively high emission efficiency either in solution or in the solid state. Up to now, a large number of D–A type compounds have been explored for red or deep red organic LEDs (OLEDs) [112,113], but rarely for NIR OLEDs. NIR emissive compounds that can be sublimated are particularly useful and desirable for device applications, because they can be obtained in high purity by sublimation and used to make a multilayer device—two important criteria for having a good device performance.

Yang et al. reported the use of two D–A–D chromophores **17** and **18** in NIR OLEDs [114]. The device with a multilayer configuration containing tris(8-hydroxyquinoline)aluminum (Alq_3) as a host and **18** as a dopant emitted the deep red light at 692 nm with a maximal EQE of 1.6%. Due to the stronger acceptor in **17**, the EL maximum of the corresponding device appeared at 815 nm (EQE = 0.5%). However, the visible emission from the Alq_3 host was still present at the concentration of 3.5% for **17** and only disappeared at the 5% doping concentration with a tradeoff of more than 25% decreased in the EQE. Recently, the same group improved the EQE of devices in nearly 2–3 folds, by using a phosphorescent sensitizer in the emissive layer of these two chromophores in order to harvest both singlet and triplet excitons formed on the host molecules [115]. The two phosphorescent emitters, $Ir(ppy)_3$ and PQIr [116,117] were used at a concentration of 10% by weight. The EQE of the sensitized NIR OLED devices reached 1.6% and 3.1% for **17** and **18**, respectively. However, these devices still emitted the strong visible light from the sensitizers (Figure 5.24), suggesting an incomplete energy transfer of triplet excitons from the phosphorescent sensitizers to the NIR fluorescent emitters. Increasing the concentration of fluorescent dopant can reduce the sensitizer emission with the decrease of EQE (Figure 5.24a).

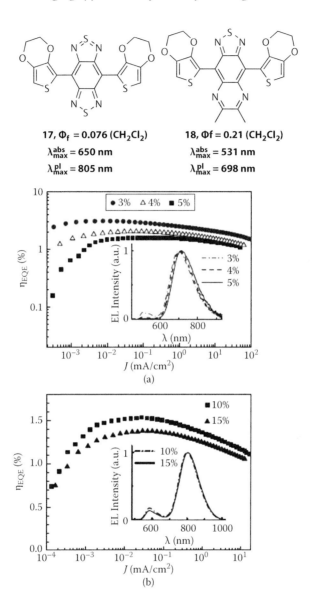

Figure 5.24 (a) EQE, η_{EQE}, as a function of the current density, J, for compound **18** based sensitized fluorescent (SF) OLEDs with different doping concentrations of NIR fluorescent emitter. Inset: normalized EL spectra for the devices in (a). (b) EQE as a function of J for compound **17** based SF devices with different doping concentrations of the phosphorescent sensitizer Inset: normalized EL spectra for the devices in (b). (Adapted with permission from Y. Yang et al., *J. Appl. Phys.* **2009**, *106*, 044509.)

19, $\Phi_f = 0.053$ (toluene)

$\lambda_{max}^{abs} = 600$ nm

$\lambda_{max}^{pl} = 752$ nm

20, $\Phi_f = 0.044$ (toluene)

$\lambda_{max}^{abs} = 620$ nm

$\lambda_{max}^{pl} = 772$ nm

21, $\Phi_f = 0.04$ (toluene)

$\lambda_{max}^{abs} = 672$ nm

$\lambda_{max}^{pl} = 830$ nm

Using the D–π–A–π–D type of NIR-fluorescent compounds is a highly feasible route to the realization of efficient NIR OLEDs in the long wavelength region. The EL properties of three thiadiazoloquinoxaline-containing compounds **19–21** were evaluated by doping in Alq$_3$ host in OLEDs [118]. All the devices showed the concentration-dependent NIR EL in a range of 700 and 1000 nm with a redshift in the emission relative to the increase in the doping concentration, attributed to a strong polarization effect of these compounds [119]. Due to the mismatch between the emission spectrum of Alq$_3$ and the absorption of dopants **20** and **21**, the visible emission was observed at a low concentration of dopant. The best performance was achieved from the device using chromophore **19**, which had the EQE of 1.12% at the wavelength around 750 nm.

22, (CH$_2$Cl$_2$)

$\lambda_{max}^{abs} = 700$ nm

$\lambda_{max}^{pl} = 1055$ nm

23, (CH$_2$Cl$_2$)

$\lambda_{max}^{abs} = 945$ nm

$\lambda_{max}^{pl} = 1285$ nm

24, $\Phi_f = 0.058$ (toluene)

$\lambda_{max}^{abs} = 723$ nm

$\lambda_{max}^{pl} = 995$ nm

25, $\Phi_f = 0.063$ (toluene)

$\lambda_{max}^{abs} = 713$ nm

$\lambda_{max}^{pl} = 970$ nm

$\lambda_{max}^{el} = 1050$ nm

26, $\Phi_f = 0.049$ (toluene)

$\lambda_{max}^{abs} = 879$ nm

$\lambda_{max}^{pl} = 1120$ nm

$\lambda_{max}^{el} \approx 1115$ nm

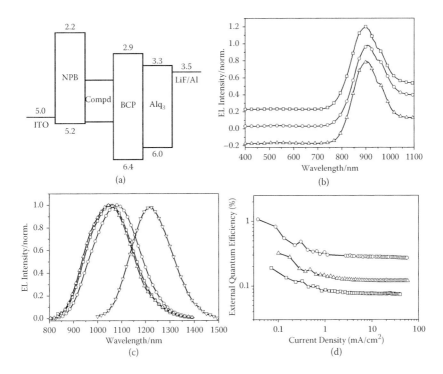

Figure 5.25 (a) Energy diagram and configuration of the OLED device. (b) Normalized EL spectra for OLEDs at 10 V recorded using the Si detector (400–900 nm). (c) Normalized NIR EL spectra for OLEDs at 10 V recorded using the InGaAs detector (800–1700 nm). (d) External quantum efficiency–current density characteristics of the devices. Compound **22** (open squares), **23** (open down triangles), **24** (open circles), and **25** (open up triangles).

Compounds **22**, **23**, **24**, and **25** have been successfully employed in NIR OLEDs with a device configuration: indium tin oxide (ITO)/MoO$_3$/1,4-bis((1-naphthylphenyl)amino)biphenyl (NPB)/emitter/2,9-dimethyl-4,7-diphenyl-1,10-phenanthroline (BCP)/Alq$_3$/LiF/Al (Figure 5.25a). Since the energy level of the emission layer falls between those of NPB and BCP, there was no emission from NPB or Alq$_3$ as the carries should be completely trapped in the emission layer (Figure 5.25b). Accordingly, the nondoped device based on **22** exhibited the exclusive NIR emission centered at 1050 nm with a maximum EQE of 0.05% (Figures 25c and 25d). Compounds **24** and **25** are structural analogs to **22** but likely have different morphology in the solid state. The nondoped devices using **24** had the EQE of 0.28% at 1080 nm, nearly four times higher than that of the device made of **22** at the same current density (Figure 5.25d). The all-organic LED device using **23** showed just one emission peak at the wavelength of 1220 nm, being the longest emission for all the known nondoped D–A type of small molecule chromophores reported to date (Figure 5.25c) [118]. Extension of the π-conjugation of spacers in compound **26** led to a redshift in absorption with a tradeoff of the aggregation-induced fluorescence quenching due to the planar geometry. The device made of compound **26** in the Alq_3 host at 2% concentration showed two emission peaks at 530 nm from Alq_3 and 1085 nm from the dopant **26**. Increasing the doping concentration to 6% could completely suppress the host emission and reveal only the NIR emission at 1115 nm [120].

The research on OLEDs based on metal complexes has recently made a steady progress in achieving long-wavelength emission and high EQE. Two classes of metal complexes have been employed as dopants or emitters in NIR OLEDs. The first class is based on trivalent lanthanide cations as the emitting centers, for example, Er^{3+} or Nd^{3+}, chelated with chromophoric ligands to sensitize excitation-energy transfer to the lanthanide ion [101]. This class of metal complex emitters is characterized by the long-wavelength emission, typically over 1000 nm and up to 1600 nm, but a rather low EQE. For example, tris(8-hydroxyquinoline)erbium, structurally similar to highly light-emitting Alq_3 was employed in OLEDs having the EL at 1500 nm and a very low EQE in the range 0.002%–0.03% [121]. Schanze et al. reported a NIR OLED utilizing Ln^{3+} in conjunction with a porphyrin/polystyrene matrix, with EQE ranging from 8.0×10^{-4} to 2.0×10^{-4}% at approximately 1 $mAcm^{-2}$ [122]. Similarly, an OLED based on $Nd(phenalenone)_3$ was reported to have a low EQE of 7×10^{-3}% at 1065 nm [123].

The second class of metal complex used in NIR OLEDs is phosphorescent transition-metal complex. They are characterized by relatively higher EQE but shorter EL wavelength (usually less than 1000 nm), in comparison with the first class of metal complexes. As the emission moves to a longer wavelength, the EQE drops significantly.

Pt(tpbp)
$\lambda_{max}^{abs} = 611$ nm
$\lambda_{max}^{pl} = 765$ nm

Pt(tptnp)
$\lambda_{max}^{abs} = 689$ nm
$\lambda_{max}^{pl} = 883$ nm

Using Pt(II)-tetraphenyltetrabenzoporphyrin [**Pt(tpbp)**] as a phospho-rescent dopant in OLED, the EQE as high as 8.5% at the wavelength of 772 nm was initially achieved, although being decreased to 5.0% at 1 mA/cm^2 [124]. Later, the same group reported the optimized OLED that emits at 765 nm with a full width at half maximum of 31 nm has a device peak EQE of 6.3% at 0.1 mAcm^{-2}, and exhibits lifetime of greater than 1000 h to 90% efficiency at 40 mAcm^{-2} [102]. The absorption spectrum of **Pt(tpbp)** displays its Q band with a maximum at 611 nm, and its phos-phorescence spectrum shows a maximal peak at 765 nm. It is known that fusing aromatic moieties at the β-pyrrole positions of porphyrin leads to a bathochromic shift of the absorption and emission energy, owing to the expansion of the π-electronic system of the porphyrin core [125]. Therefore, **Pt(tptnp)** (tptnp = tetraphenyltetranaphtho[2,3]porphyrin) has a lower energy gap than **Pt(tpbp)** and its absorption spectrum exhibits a single Q band at 689 nm. The PL of **Pt(tptnp)** is dominated by a single phosphorescence band (with a quantum yield of 0.22) with a maximal peak at 883 nm. Thus, the two different OLEDs based on the **Pt(tptnp)** phos-phor were fabricated by vacuum thermal evaporation and compared to probe the effect of addition of Cs dopant in the electron transport layer (ETL) [126]. The emissive layer (EML), consisting of 4,4′-bis(carbazol-9-yl) biphenyl mixed with 8 wt% **Pt(tptnp)**, was sandwiched between a hole transport layer (HTL) of bis[*N*-(1-naphthyl)-*N*-phenylamino]biphenyl (R-NPD) and an ETL of 4,7-diphenyl-1,10-phenanthroline (BPhen). In the so-called undoped device, the entire 100 nm ETL layer is nominally undoped without adding any other dopant. In another so-called n-doped device, Cs was used to *n*-dope BPhen (BPhen:Cs = 1:0.2, molar ratio) in order to increase the conductivity of the ETL and improve the efficiency of elec-tron injection from the cathode [127]. Both devices emit the light exclu-sively at 896 nm at low turn-on voltage (~2 V). The *R-J-V* characteristics of the doped and undoped OLEDs are compared in Figure 5.26. At $V > 2.5$ V, the current density was approximately 2 orders of magnitude higher in the *n*-doped device, due to the significantly enhanced conductivity in the *n*-doped ETL compared with that in the nominally undoped ETL. Maximum radiant emittances (*R*) of 1.8 mW/cm^2 were obtained at 17 V in the undoped device and at 12 V in the n-doped device. For the undoped device, the EQE is relatively constant at low current densities and reaches a maximum of 3.8% at $J ≈ 0.1$ mA/cm^2; however, it decreases significantly with the increase of the current density to 2.0% at $J = 10$ mA/cm^2 and 0.6% at $J = 100$ mA/cm^2. The significant roll-off in efficiency at higher current densities is likely due to the triplet–triplet exciton annihilation process that commonly occurs in phosphorescent OLEDs. The maximum power efficiency ($η_p$) of the undoped device is 19 mW/W at low current densities ($J ≈ 10^{-3}$ mA/cm^2). In comparison with the undoped device, the *n*-doped device has slightly lower quantum efficiencies with a maximum of EQE

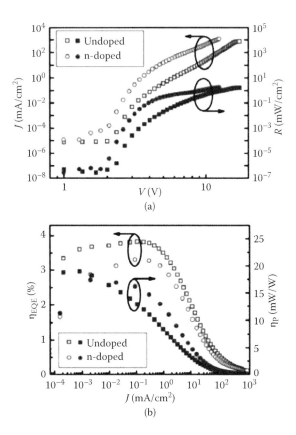

Figure 5.26 Device characteristics for undoped and n-doped OLEDs based on **Pt(tptnp)**: (a) current density, J, and the radiant emittance in the forward viewing directions, R, as functions of the voltage, V, for the undoped and n-doped devices; (b) external quantum efficiency, η_{EQE}, and power efficiency, η_p, of these two devices as functions of J. (Adapted with permission from J. R. Sommer et al., *ACS Appl. Mater. Inter.* **2009**, *1*, 274. Copyright (2009) American Chemical Society.)

of 3.3%. At $J = 1$ mA/cm^2, $\eta_p = 12$ mW/W for the n-doped device, which is more than 40% higher than that of the undoped device (8.4 mW/W).

Complexes containing other metal ions have also been explored as NIR emitters. For example, the OLED that used a cyclometalated [(pyrenyl–quinolyl)$_2$Ir(acac)] complex as the phosphor emitted at 720 nm with the EQE of 0.1% [128]. The electroluminescent devices employing 6 wt% of a charge neutral Os(ibtz)$_2$(PPhMe$_2$)$_2$ [ibtz = 5-(1-isoquinolyl)-3-*tert*-butyl-1,2,4-triazole] doped in the Alq$_3$ host exhibited fairly intense emission with a peak wavelength at 814 nm. The devices had the forward radiant emittance as high as 65 µWcm^{-2} and a peak EQE of ~1.5% [129].

Despite of all these progress made so far in the field of NIR OLED, it is still necessary and desirable to continue exploring the scope and limitations of various types of highly efficient NIR emitters. Further materials and device optimization, such as tuning of material morphology or use of different host materials and assistant dopants for energy cascade transfer, are likely to improve the overall performance of NIR OLEDs. The successful development of NIR OLED will realize completely new niche applications for the already commercially successful LED technology.

In 2012, at the Smart Systems Integrations Exhibition in Zurich, the Center for Organic Materials and Electronic Devices of the Fraunhofer Institute for Photonic Microsystems in Dresden, Germany, showcased a demonstrator of NIR OLEDs integrated in CMOS-silicon chips. Integration of NIR OLEDs in CMOS-silicon chips allows for many niche applications, such as a light source in the field of sensor technology, for medical light therapy or as a display.

The showcased NIR OLED contains phosphorescent small molecules and emits the light at the maximal wavelength of ~770 nm (Figure 5.27a). By using the pin-OLED technology, the external power conversion efficiency of 2.5% could be achieved at a low operating voltage (3.5 V at 10 mA/cm^2). After 950 h, the initial luminance decreased only by 10% or lifetime (LT90) of 950 h at 100 mA/cm^2. For comparison, if an orange pin-OLED is driven at 100 mA/cm^2, it reaches an LT50 of 161 h. Therefore, with respect to the lifetime, the NIR OLED is able to compete with the visibly colored analogs.

The first, polychromatic, bidirectional OLED microdisplay has been developed with a local emission in the NIR regime. As shown in Figure 5.27b, an OLED microdisplay with 0.6″ screen diagonal and QVGA-display-resolution (320 × 240 pixel) and integrated camera (160 × 120 pixel) is divided in four segments and deposited with blue, green, red, and NIR emitting (730–850 nm) organic layers. The NIR segment remains invisible for the observer (Figure 5.27b, right bottom), while the NIR emission can only be detected with a camera without a NIR filter (Figure 5.27c).

By combining of NIR-OLEDs with conventional OLEDs, light systems with an accordingly extended spectral range can be produced for color sensors or spectrometers. NIR displays could be applied, for example, in night-vision devices fitted with supporting low-light amplifiers. The displayed information would remain invisible to anyone but the operator of the device. Furthermore, the new technology is able to produce bidirectional NIR OLED displays that cannot only emit NIR-light, but also detect it.

Imaging techniques are a vital part of clinical diagnostics, biomedical research, and nanotechnology. Optical molecular imaging makes use

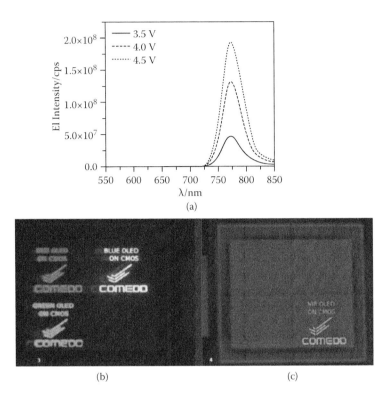

Figure 5.27 (See color insert.) (a) EL spectra of an NIR OLED at different applied bias. (b) Detailed image of an RGB-NIR microdisplay recorded in the visible range of light. (c) Detailed image of RGB-NIR microdisplay recorded in the NIR region of light.

of relatively harmless, low-energy light and technically straightforward instrumentation. Chemiluminescence (CL) has emerged to be an extremely useful detection technology in clinical diagnostics both for immuno-assays and nucleic acid assays [130]. CL has also been utilized as a sensitive and convenient analytical tool in food analysis [131], environmental and forensic sciences [132], and for detection of hydrogen peroxide and biologically active compounds (e.g., glucose, folic acid, or lactic acid) [133]. For biological imaging, self-illuminating chemiluminescent probes have a signal contrast advantage over fluorescent probes because there is no excitation beam and thus no background autofluorescence by endogenous biomolecules [134]. Currently, CL imaging involves short-lived molecular species that must be generated in situ. Most of the currently available chemiluminescent probes decay by emitting visible light, which is readily absorbed and largely scattered by the molecules and cells in biological matrices, and thus does not penetrate far through heterogeneous biological media [135]. Therefore, an urgent need for next-generation CL systems that emit light within the optimal wavelength window of 700–900 nm or even 900–1200 nm is well recognized [136].

CL is the emission of energy and differs from photoluminescence in that the electronic excited state is derived from the product of chemical reaction rather than absorption as a more typical way of creating electronic excited states. Although the light-emitting chemical reactions have been widely observed [137], only a few chemiluminescent reactions are efficient and useful for practical applications. Typical examples include the reactions of an oxidizing agent (e.g., hydrogen peroxide or oxygen) with phthalhydrazide derivatives (luminol) [138], dimethylbiacridinium ion (lucigenin) [139], imidazole derivatives (lophine) [140], acridinium salts [141], and oxalyl chloride in the presence of fluorescent dyes [142]. Chemiluminescent acridinium ester labels with the emission in the visible region (e.g., 420–520 nm) are widely used in clinical diagnostics especially in automated immunochemistry analyzers such as Siemens Healthcare Diagnostics' ADVIA Centaur systems [143]. Structural modification aiming at extension of the CL wavelength into the NIR region only led to acridinium esters with the PL and CL emission wavelengths close to 700 nm [144].

A new paradigm for optical molecular imaging was reported in 2010 [145]. The CL source is based on a squaraine rotaxane endoperoxide (SREP), which is a permanently interlocked [2] rotaxane comprising a dumbbell-shaped squaraine dye encapsulated by a tetralactam macrocycle that contains a thermally unstable 9,10-anthracene endoperoxide group [146]. Squaraine rotaxane endoperoxides can be stored indefinitely at temperatures below –20°C, but upon warming to body temperature, they undergo a unimolecular chemical reaction and emit NIR light (λ_{max} ~ 730 nm) that can pass through a living mouse. As shown in Figure 5.28, in a typical experimental setup for planar CL, NIR CL from a small tube containing a solution of SREP (250 nmol) passes through a living nude mouse positioned between the tube and the CCD camera. The target background ratio for the transmitted light is 11.6, which is quite impressive in comparison with the value of 1.1 obtained from the same experimental arrangement plus excitation light for planar fluorescence. Signal intensity is the strongest at each side of the animal, which coincides with passage through the least amount of tissue.

At present, most of CL luminophores produce the visible emission and a very few NIR fluorescent dyes are known to exhibit rather weak NIR CL [147]. Owing to a large number of recently available NIR fluorescent D–A compounds and polymers, it is highly feasible to develop new NIR CL with emission above 1000 nm. In 2012, Wang et al. reported, for the first time, NIR CL tunable from 900 nm to 1700 nm from the reaction of low

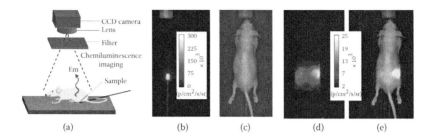

Figure 5.28 (See color insert.) Chemiluminescence from squaraine rotaxane endoperoxide (SREP) at 38°C penetrates through a living nude mouse. (a) Experimental setup for planar chemiluminescence. (b) Chemiluminescence pixel intensities from a small tube containing SREP (250 nmol) in $C_2D_2Cl_4$. (c) Photographs of mouse located above the tube. (d,e) Pixel intensity map of chemiluminescence that is transmitted through the mouse (target background ratio = 11.6). (Adapted with permission from J. M. Baumes et al., *Nature Chem.* **2010**, *2*, 1025.)

band-gap compounds and polymers with oxalyl chloride and hydrogen peroxide [148]. Prolonged chemiluminescent process in several hundreds of seconds can be achieved in solutions at low temperature or in polymer gels at room temperature. Therefore, the luminophore–polymer gel system is a useful source for producing the prolonged NIR light for possible medical applications.

References

1. C. Xu, J. Ye, D. L. Marks, S. A. Boppart, *Opt. Lett.* **2004**, *29*, 1647.
2. N. Cauchon, H. Tian, R. Langlois, C. La Madeleine, S. Martin, H. Ali, D. Hunting, J. E. van Lier, *Bioconjugate Chem.* **2005**, *16*, 80.
3. C. Comuzzi, S. Cogoi, M. Overhand, G. A. Van der Marel, H. S. Overkleeft, L. E. Xodo, *J. Med. Chem.* **2006**, *49*, 196.
4. (a) J. Fabian, H. Nakazumi, M. Matsuoka, *Chem. Rev.* **1992**, *92*, 1197. (b) M. Matsuoka, Ed. *Infrared Absorbing Dyes*; Plenum: New York, 1990.
5. (a) M. Yousaf, M. Lazzouni, *Dyes Pigm.* **1995**, *27*, 297. (b) F. D. Robert, *Development of New Laser Protective Dyes.* **1993**, *AD-A*, 260502.
6. (a) T. Shimomura, S. Onomichi, M. Kobayashi, Y. Yamada, S, Yokoyama, US Patent 6,522,463 (February 18, 2003). (b) A. Ito, S. Ishida, T. Iida, T. Ishii, N. Kobayashi, US Patent 7,887,920 B2 (Feb. 15, 2011).
7. G. M. Fischer, E. Daltrozzo, A. Zumbusch, *Angew. Chem. Int. Ed.* **2011**, *50*, 1.
8. I. Jones, R. Wise, WO 00/20157.
9. S. D. DeCoste, W. Farinelli, T. Flotte, R. R. Anderson, *Lasers Surg. Med.* **1992**, *12*, 25.
10. U. T. Müller-Westerhof, B. Vance, D. I. Yoon, *Tetrahedron*, **1991**, *47*, 909.
11. M. C. Aragoni, M. Arca, T. Cassano, C. Denotti, F. A. Devillanova, F. Isaia, V. Lippolis, D. Natali, L. Nitti, M. Sampietro, R. Tommasi, G. Verani, *Inorg. Chem. Commun.* **2002**, *5*, 869.

12. B. Barber, C. R. Giles, V. Askyuk, R. Ruel, L. Stulzl, D. Bishop, *IEEE Photonics Techno. Lett.* **1998**, *10*, 1262.
13. T. Kawai, M. Koga, M. Okuno, T. Kitoh, *Electron. Lett.* **1998**, *34*, 264.
14. Y.-S. Jin, S.-S. Lee, Y.-S. Son, *Electron. Lett.* **1998**, *35*, 916.
15. (a) N. A. O'Brien, E. R. Mathew, B. P. Hichwa, *OFC 1999*, PD26-1. (b) M. A. Cowin, R. Varrazza, C. Morgan, R. V. Penty, I. H. White, A. M. McDonagh, S. Bayly, J. Riley, M. D. Ward, J. A. McCleverty, *OFC 2001*, WR6-1. (c) N. C. Harden, E. R. Humphrey, J. C. Jeffery, S. M. Lee, M. Marcaccio, J. A. McCleverty, L. H. Rees, M. D. Ward, *J. Chem. Soc., Dalton Trans.* **1999**, 2417. (d) M. D. Ward, *J. Solid State Electrochem.* **2005**, *9*, 778.
16. M. D. Ward, *Chem. Soc. Rev.* **1995**, *24*, 121.
17. (a) Y. Qi, P. Desjardins, Z. Y. Wang, *J. Opt. A. Pure Appl. Opt.* **2002**, *4*, S273. (b) Y. Qi, P. Desjardins, Z. Y. Wang, *Opt. Mater.* **2002**, *21*, 255. (c) Y. Qi, Z. Y. Wang, *Macromolecules*, **2003**, *36*, 3146.
18. B. Sullivan, D. Salmon, T. Meyer, *Inorg. Chem.* **1978**, *17*, 3334.
19. G. LeClair, Z. Y. Wang, *J. Solid State Electrochem.* **2009**, *13*, 365.
20. J. Zhang, X. Wu, H. Yu, D. Yan, Z. Y. Wang, *Chin. Sci. Bull.* **2005**, *50*, 2688.
21. C. D. MacPherson, K. Anderson, S. P. McGarry, US Patent 2002/0067905A1 (June 6, 2002).
22. (a) I. Schwendeman, J. Hwang, D.M. Welsh, D.B. Tanner, J.R. Reynolds, *Adv. Mater.* **2001**, *13*, 634. (b) H. Meng, D. Tucker, S. Chaffins, Y. Chen, R. Helgeson, B. Dunn, F. Wudl, *Adv. Mater.* **2003**, *15*, 146.
23. J. D. Zhang, H. A. Yu, X. G. Wu, D. H. Yan, Z. Y. Wang, *Opt. Mater.* **2004**, *27*, 265.
24. B. P. Jelle, G. Hagen, *Solar Energy Mater. Solar Cells*, **1999**, *58*, 277.
25. M. Biancardo, P. F. H. Schwab, R. Argazzi, C. A. Bignozzi, *Inorg. Chem.* **2003**, *42*, 3966.
26. (a) R. J. Mortimer, N. M. Rowley, in *Comprehensive Coordination Chemistry*, M. D. Ward (Ed.), 2nd ed., Vol. 9. Elsevier, Oxford, pp. 581–619 (2003). (b) M. C. Aragoni, M. Arca, F. Demartin, F. A. Devillanova, A. Garau, F. Isaia, F. Lelj, V. Lippolis, G. Verani, *J. Am. Chem. Soc.* **1999**, *121*, 7098. (c) F. Bigoli, P. Cassoux, P. Deplano, M. L. Mercuri, M. A. Pellinghelli, G. Pintus, A. Serpe, E. F. Trogu, *J. Chem. Soc. Dalton Trans.* **2000**, 4639.
27. S. Xun, G. LeClair, J. Zhang, X. Chen, J. P. Gao, Z. Y. Wang, *Org. Lett.* **2006**, *8*, 1697.
28. (a) P. W. Barone, S. Baik, D. A. Heller, M. S. Strano, *Nature Mater.* **2005**, *4*, 86. (b) P. S. Jensen, J. Bak, S. Andersson-Engels, *Appl. Spectrosc.* **2003**, *57*, 28.
29. B. O'Regan, M. Grätzel, *Nature*, **1991**, *353*, 737.
30. (a) M. Grätzel, *Nature*, **2001**, *414*, 338. (b) A. Hagfeldt, M. Grätzel, *Acc. Chem. Rec.* **2000**, *33*, 269.
31. S. Ardo, G. J. Meyer, *Chem. Soc. Rev.* **2009**, *38*, 115.
32. A. Hagfeldt, G. Boschloo, L. Sun, L. Kloo, H. Pettersson, *Chem. Rev.* **2010**, *110*, 6595.
33. A. Yella, H.-W. Lee, H. N. Tsao, C. Yi, A. K. Chandiran, Md. K. Nazeeruddin, E. W.-G. Diau, C.-Y. Yeh, S. M Zakeeruddin, M. Grätzel, *Science*, **2011**, *334*, 629.
34. B. E. Hardin1, H. J. Snaith, M. D. McGehee, *Nature Photon.* **2012**, *6*, 162.
35. C.-Y. Chen, M. Wang, J.-Y. Li, N. Pootrakulchote, L. Alibabaei, C. Ngoc-le, J.-D. Decoppet, J.-H. Tsai, C. Grätzel, C.-G. Wu, S. M. Zakeeruddin, M. Grätzel, *ACS Nano*, **2009**, *3*, 3103.

36. J. Burschka, A. Dualeh, F. Kessler, E. Baranoff, N. L. Cevey-Ha, C. Yi, M. K. Nazeeruddin, M. Grätzel, *J. Am. Chem. Soc.* **2011**, *133*, 18042.
37. (a) H. J. Snaith, *Adv. Funct. Mater.* **2010**, *20*, 13. (b) T. W. Hamann, R. A. Jensen, A. B. F. Martinson, H. V. Ryswyk, J. T. Hupp, *Energ. Environ. Sci.* **2008**, *1*, 66.
38. (a) T. Ono, T. Yamaguchi, H. Arakawa, *Sol. Energ. Mater. Sol. C*, **2009**, *93*, 831. (b) L. Macor, F. Fungo, T. Tempesti, E. N. Durantini, L. Otero, E. M. Barea, F. Fabregat-Santiago, J. Bisquert, *Energ. Environ. Sci.* **2009**, *2*, 529. (c) T. Maeda, K. Miyanaga, N. Shima, S. Yagi, Y. Hamamura, H. Nakazumi, *Org. Lett.* **2011**, *13*, 5994.
39. S. Ito, S. M. Zakeeruddin, R. Humphry-Baker, P. Liska, R. Charvet, P. Comte, M. K. Nazeeruddin, P. Péchy, M. Takata, H. Miura, S. Uchida, M. Grätzel, *Adv. Mater.* **2006**, *18*, 1202.
40. Y. Jin, J. Hua, W. Wu, X. Ma, F. Meng, *Synth. Met.* **2008**, *158*, 64.
41. C. C. Lenznoff, A. B. P. Lever, in *Phthalocyanines,* Vol. 3, VCH: New York, 1993.
42. J.-H. Yum, P. Walter, S. Huber, D. Rentsch, T. Geiger, F. Nuesch, F. De Angelis, M. Gratzel, M. K. Nazeeruddin, *J. Am. Chem. Soc.* **2007**, *129*, 10320.
43. G. K. Mor, S. Kim, M. Paulose, O. K. Varghese, K. Shankar, J. Basham, C. A. Grimes, *Nano Lett.* **2009**, *9*, 4250.
44. A. Burke, L. Schmidt-Mende, S. Ito, M. Grätzel, *Chem. Commun.* **2007**, 234.
45. Y. Hao, X. Yang, J. Cong, H. Tian, A. Hagfeldt, L. Sun, *Chem. Commun.* **2009**, 4031.
46. B. E. Hardin, A. Sellinger, T. Moehl, R. Humphry-Baker, J.-E. Moser, P. Wang, S. M. Zakeeruddin, M. Grätzel, M. D. McGehee, *J. Am. Chem. Soc.* **2011**, *133*, 10662.
47. C. W. Tang, *Appl. Phys. Lett.* **1986**, *48*, 183.
48. C. J. Brabec, *Sol. Energy Mater. Sol. Cells*, **2004**, *83*, 273.
49. (a) J. Y. Kim, K. Lee, N. E. Coates, D. Moses, T.-Q. Nguyen, M. Dante, A. J. Heeger, *Science*, **2007**, *317*, 222. (b) E. Bundgaard, F. C. Krebs, *Sol. Energy Mater. Sol. Cells*, **2007**, *91*, 954.
50. C. Winder, N. S. Sariciftci, *J. Mater. Chem.* **2004**, *14*, 1077.
51. (a) C. J. Brabec, N. S. Sariciftci, J. C. Hummelen, *Adv. Funct. Mater.* **2001**, *11*, 15. (b) J. Hou, H.-Y. Chen, S. Zhang, G. Li, Y. Yang, *J. Am. Chem. Soc.* **2008**, *130*, 16144. (c) Y. Liang, Y. Wu, D. Feng, S.-T. Tsai, H.-J. Son, G. Li, L. Yu, *J. Am. Chem. Soc.* **2009**, *131*, 56.
52. H. Hoppe, N. S. Sariciftci, *J. Mater. Chem.* **2006**, *16*, 45.
53. M. Sun, L. Wang, X. Zhu, B. Du, R. Liu, W. Yang, Y. Cao, *Sol. Energy Mater. Sol. Cells*, **2007**, *91*, 1681.
54. A. B. Tamayo, B. Walker, T.-Q. Nguyen, *J. Phys. Chem. C*, **2008**, *112*, 11545.
55. A. B. Tamayo, X.-D. Dang, B. Walker, J. Seo, T. Kent, T.-Q. Nguyen, *Appl. Phys. Lett.* **2009**, *94*, 103301.
56. J. Peet, A. B. Tamayo, X. D. Dang, J. H. Seo, T. Q. Nguyen, *Appl. Phys. Lett.* **2008**, *93*, 163306.
57. M. C. Scharber, D. Mühlbacher, M. Koppe, P. Denk, C. Waldauf, A. J. Heeger, C. J. Brabec, *Adv. Mater.* **2006**, *18*, 789.
58. B. Walker, A. B. Tamayo, X.-D. Dang, P. Zalar, J. H. Seo, A. Garcia, M. Tantiwiwat, T.-Q. Nguyen, *Adv. Funct. Mater.* **2009**, *19*, 3063.
59. (a) N. S. Sariciftci, L. Smilowitz, A. J. Heeger, F. Wudl, *Science*, **1992**, *258*, 1474. (b) S. Morita, A. A. Zakhidov, K. Yoshino, *Solid State Commun.* **1992**, *82*, 249.

60. (a) G. Yu, J. Gao, J. C. Hummelen, F. Wudl, A. J. Heeger, *Science*, **1995**, *270*, 1789. (b) J. J. M. Halls, C. A. Walsh, N. C. Greenham, E. A. Marseglia, R. H. Friend, S. C. Moratti, A. B. Holmes, *Nature*, **1995**, *376*, 498.

61. (a) G. Li, V. Shrotriyai, J. Huang, Y. Yao, T. Moriarty, K. Emery, Y. Yang, *Nature Mater.* **2005**, *4*, 864. (b) W. L. Ma, C. Y. Yang, X. Gong, K. Lee, A. J. Heeger, *Adv. Funct. Mater.* **2005**, *15*, 1617.

62. (a) C. J. Brabec, S. E. Shaheen, C. Winder, N. S. Sariciftci, P. Denk, *Appl. Phys. Lett.* **2002**, *80*, 1288. (b) M. M. Wienk, J. M. Kroon, W. J. H. Verhees, J. Knol, J. C. Hummelen, P. A. van Hal, R. A. J. Janssen, *Angew. Chem. Int. Ed.* **2003**, *42*, 3371.

63. F. Padinger, R. S. Rittberger, N. S. Sariciftci, *Adv. Funct. Mater.* **2003**, *13*, 85.

64. Z. Bao, A. Dodabalapur, A. Lovinger, *Appl. Phys. Lett.* **1996**, *69*, 4108.

65. (a) Y. J. Cheng, S. H. Yang, C. S. Hsu, *Chem. Rev.* **2009**, *109*, 5868. (b) Y. Y. Liang, L. P. Yu, *Polym. Rev.* **2010**, *50*, 454.

66. Y. Y. Liang, L. P. Yu, *Acc. Chem. Res.* **2010**, *43*, 1227.

67. (a) E. E. Havinga, W. Tenhoeve, H. Wynberg, *Synth. Met.* **1993**, *55*, 299. (b) Q. T. Zhang, J. M. Tour, *J. Am. Chem. Soc.* **1997**, *119*, 5065.

68. D. Muhlbacher, M. Scharber, M. Morana, Z. G. Zhu, D. Waller, R. Gaudiana, C. Brabec, *Adv. Mater.* **2006**, *18*, 2884.

69. J. Peet, J. Y. Kim, N. E. Coates, W. L. Ma, D. Moses, A. J. Heeger, G. C. Bazan, *Nature Mater.* **2007**, *6*, 497.

70. N. Blouin, A. Michaud, M. Leclerc, *Adv. Mater.* **2007**, *19*, 2295.

71. S. H. Park, A. Roy, S. Beaupre, S. Cho, N. Coates, J. S. Moon, D. Moses, M. Leclerc, K. Lee, A. J. Heeger, *Nature Photon.* **2009**, *3*, 297.

72. F. Huang, K.-S. Chen, H.-L. Yip, S. K. Hau, O. Acton, Y. Zhang, J. Luo, A. K. Y. Jen, *J. Am. Chem. Soc.* **2009**, *131*, 13886.

73. (a) Y. Y. Liang, D. Q. Feng, Y.Wu, S. T. Tsai, G. Li, C. Ray, L. P. Yu, *J. Am. Chem. Soc.* **2009**, *131*, 7792. (b) H. Y. Chen, J. Hou, S. Zhang, Y. Liang, G. Yang, Y. Yang, L. Yu, Y. Wu, G. Li, *Nature Photon.* **2009**, *3*, 649.

74. (a) S. C. Price, A. C. Stuart, L. Q. Yang, H. X. Zhou, W. You, *J. Am. Chem. Soc.* **2011**, *133*, 4625. (b) H. Zhou, L. Yang, A. C. Stuart, S. C. Price, S. Liu, W. You, *Angew. Chem. Int. Ed.* **2011**, *50*, 2995. (c) M.-S. Su, C.-Y. Kuo, M.-C. Yuan, U-S. Jeng, C.-J. Su, K.-H. Wei, *Adv. Mater.* **2011**, *23*, 3315. (d) J. Yang, R. Zhu, Z. R. Hong, Y. J. He, A. Kumar, Y. F. Li, Y. Yang, *Adv. Mater.* **2011**, *23*, 3465. (e) Y. Sun, C. J. Takacs, S. R. Cowan, J. Hwa Seo, X. Gong, A. Roy, A. J. Heeger, *Adv. Mater.* **2011**, *23*, 2226. (f) T.-Y. Chu, J. P. Lu, S. Beaupr, Y. Zhang, J.-R. Pouliot, S. Wakim, J. Zhou, M. Leclerc, Z. Li, J. Ding, Y. Tao, *J. Am. Chem. Soc.* **2011**, *133*, 4250. (g) C. M. Amb, S. Chen, K. R. Graham, J. Subbiah, C. E. Small, F. So, J. R. Reynolds, *J. Am. Chem. Soc.* **2011**, *133*, 10062.

75. H. Y. Chen, J. Hou, A. E. Hayden, H. Yang, K. N. Houk, Y. Yang, *Adv. Mater.* **2010**, *22*, 371.

76. C. Piliego, T.W. Holcombe, J. D. Douglas, C. H. Woo, P. M. Beaujuge, J. M. J. Fréchet, *J. Am. Chem. Soc.* **2010**, *132*, 7595.

77. Y. Xia, L. Wang, X. Deng, D. Li, X. Zhu, Y. Cao, *Appl. Phys. Lett.* **2006**, *89*, 081106.

78. X. Gong, M. Tong, Y. Xia, W. Cai, J. S. Moon, Y. Cao, G. Yu, C.-L. Shieh, B. Nilsson, A. J. Heeger, *Science*, **2009**, *325*, 1665.

79. (a) A. Rogalski, J. Antoszewski, L. Faraone, *J. Appl. Phys.* **2009**, *105*, 091101. (b) M. Ettenberg, *Adv. Imaging*, **2005**, *20*, 29. (c) E. H. Sargent, *Adv. Mater.* **2005**, *17*, 515. (d) S. Kim, Y. T. Lim, E. G. Soltesz, A. M. De Grand, J. Lee, A. Nakayama, J. A Parker, T. Mihaljevic, R. G. Laurence, D. M. Dor, L. H. Cohn,

M. G. Bawendi, J. V. Frangioni, *Nature Biotechnol.* **2004**, *22*, 93. (e) A. R. Jha, *Infrared Technology Applications to Electrooptics, Photonic Devices, and Sensors*, Wiley: New York **2000**, pp. 245–438. (f) H. C. Liu, J.-P. Noel, Lujian Li, M. Buchanan, J. G. Simmons, *Appl. Phys. Lett.* **1992**, *60*, 3298. (g) K. K. Choi, G. Dang, J. W. Little, K. M. Leung, T. Tamir, *Appl. Phys. Lett.* **2004**, *84*, 4439. (h) K. Stewart, M. Buda, J. Wong-Leung, L. Fu, C. Jagadish, A. Stiff-Roberts, P. Bhattacharya, *J. Appl. Phys.* **2003**, *94*, 5283. (i) S. Krishna, S. Raghavan, G. von Winckel, P. Rotella, A. Stintz, C. P. Morath, D. Le, S. W. Kennerly, *Appl. Phys. Lett.* **2003**, *82*, 2574. (j) A. Rogalski, *Infrared Phys. Technol.* **2002**, *43*, 187.

80. Y. Yao, Y. Liang, V. Shrotriya, S. Xiao, L. Yu, Y. Yang, *Adv. Mater.* **2007**, *19*, 3979.

81. G. Qian, J. Qi, J. A. Davey, J. S. Wright, Z. Y. Wang, *Chem. Mater.* **2012**, *24*, 2364.

82. J. D. Zimmerman, E. K. Yu, V. V. Diev, K. Hanson, M. E. Thompson, S. R. Forrest, *Org. Electron.* **2011**, *12*, 869.

83. P. V. V. Jayaweera, A. G. U. Perera, *Appl. Phys. Lett.* **2004**, *85*, 5754.

84. L. A. DeWerd, P. R. Moran, *Med. Phys.* **1978**, *5*, 23.

85. (a) K. Y. Law, *Chem. Rev.* **1993**, *93*, 449. (b) D. S. Weiss, M. Abkowitz, *Chem. Rev.* **2010**, *110*, 479. (c) S. Grammatica, J. Mort, *Appl. Phys. Lett.* **1981**, *38*, 445.

86. W. West, P. B. Gilman, in *The Theory of the Photographic Process*, 4th ed., T. H. James, Ed., Macmillan: New York, 1977, Chapter 10.

87. S. Kazaoui, N. Minami, B. Nalini, Y. Kim, *Appl. Phys. Lett.* **2005**, *98*, 084314.

88. (a) O. Korovyanko, C. Sheng, Z. Vardeny, A. Dalton, R. Baughman, *Phys. Rev. Lett.* **2004**, *92*, 017403. (b) F. Wang, G. Dukovic, L. Brus, T. Heinz, *Science*, **2005**, *308*, 838. (c) V. Perebeinos, J. Tersoff, P. Avouris, *Phys. Rev. Lett.* **2005**, *94*, 027402. (d) C. D. Spataru, S. Ismail-Beigi, L. X. Benedict, S. G. Louie, *Phys. Rev. Lett.* **2004**, *92*, 077402.

89. M. Freitag, Y. Martin, J. A. Misewich, R. Martel, P. Avouris, *Nano Lett.* **2003**, *3*, 1067.

90. F. Bigoli, P. Deplano, F. A. Devillanova, J. R. Ferraro, V. Lippolis, P. J. Lukes, M. L. Mercuri, M. A. Pellinghelli, E. F. Trogu, J. M. Williams, *Inorg. Chem.* **1997**, *36*, 1218.

91. M. C. Aragoni, M. Arca, T. Cassano, C. Denotti, F. A. Devillanova, F. Isaia, V. Lippolis, D. Natali, L. Nitti, M. Sampietro, R. Tommasi, G. Verani, *Inorg. Chem. Commun.* **2002**, *5*, 869.

92. (a) H. Usta, A. Facchetti, T. J. Marks, *J. Am. Chem. Soc.* **2008**, *130*, 8580. (b) M. Shkunov, R. Simms, M. Heeney, S. Tierney, I. McCulloch, *Adv. Mater.* **2005**, *17*, 2608. (c) A. Babel, Y. Zhu, K. F. Cheng, W. C. Chen, S. A. Jenekhe, *Adv. Funct. Mater.* **2007**, *1*, 2542. (d) E. J. Meijer, D. M. Deleeuw, S. Setayesh, E. Van Veenendaal, B.-H. Huisman, P. W. M. Blom, J. C. Hummelen, U. Scherf, T. M. Klapwijk, *Nature Mater.* **2003**, *2*, 678. (e) J. Cornil, J.-L. Brédas, J. Zaumseil, H. Sirringhaus, *Adv. Mater.* **2007**, *19*, 1791. (f) K. Szendrei, D. Jarzab, Z. Chen, A. Facchetti, M. A. Loi, *J. Mater. Chem.* **2010**, *20*, 1317.

93. Y. Li, S. P. Singh, P. Sonar, *Adv. Mater.* **2010**, *22*, 4862.

94. P. Sonar, S. P. Singh, Y. Li, M. S. Soh, A. Dodabalapur, *Adv. Mater.* **2010**, *22*, 5409.

95. J. Fan, J. D. Yuen, M. Wang, J. Seifter, J.-H. Seo, A. R. Mohebbi, D. Zakhidov, A. Heeger, F. Wudl, *Adv. Mater.* **2012**, *24*, 2186.

96. G. Patonay, J. Salon, J. Sowell, L. Strekowski, *Molecules*, **2004**, *9*, 40.

97. C.-T. Chen, S. R. Marder, *Adv. Mater.* **1995**, *7*, 1030.

98. H. T. Whelan, R. L. Smits, E. V. Buchman, N. T. Whelan, S. G. Turner, D. A. Margolis, V. Cevenini, H. Stinson, R. Ignatius, T. Martin, J. Cwiklinski, A. F. Philippi, W. R. Graf, B. Hodgson, L. Gould, M. Kane, G. Chen, J. Caviness, *J. Clin. Laser Med. Surg.* **2001**, *19*, 305.

99. R. Raghavachari, *Near-Infrared Applications in Biotechnology*, CRC Press: Boca Raton, FL, 2001.

100. E. Desurvire, *Erbium-Doped Fiber Amplifiers: Principles and Applications*, Wiley-Interscience: New York, 1994.

101. J. Kido, Y. Okamoto, *Chem. Rev.* **2002**, *102*, 2357.

102. C. Borek, K. Hanson, P. I. Djurovich, M. E. Thompson, K. Aznavour, R. Bau, Y. Sun, S. R. Forrest, J. Brooks, J. Michalski, J. Brown, *Angew. Chem. Int. Ed.* **2007**, *46*, 1109.

103. M. Chen, E. Perzon, M. R. Andersson, S. Marcinkevicius, S. K. M. Jonsson, M. Fahlman, M. Berggren, *Appl. Phys. Lett.* **2004**, *84*, 3570.

104. A. Fujii, M. Yoshida, Y. Ohmori, K. Yoshino, *Jpn. J. Appl. Phys. Part 2.* **1996**, *35*, L37.

105. A. Fujii, Y. Ohmori, K. Yoshino, *IEEE Trans. Electron Dev.* **1997**, *44*, 1204.

106. H. Suzuki, *Appl. Phys. Lett.* **2000**, *76*, 1543.

107. H. Suzuki, *Appl. Phys. Lett.* **2002**, *80*, 3256.

108. H. Suzuki, K. Ogura, N. Matsumoto, P. Prosposito, S. Schutzmann, *Mol. Cryst. Liq. Cryst.* **2006**, *444*, 51.

109. E. I. Maltsev, D. A. Lypenko, V. V. Bobinkin, A. R. Tameev, S. V. Kirillov, B. I. Shapiro, H. F. M. Schoo, A. V. Vannikov, *Appl. Phys. Lett.* **2002**, *81*, 3088.

110. Y. Xuan, G. Qian, Z. Wang, D. Ma, *Thin Solid Films.* **2008**, *516*, 7891.

111. Y. Xuan, N. Zhao, D. Pan, X. Ji, Z. Wang, D. Ma, *Semicond. Sci. Technol.* **2007**, *22*, 1021.

112. B.-J. Jung, C.-B. Yoon, H.-K. Shim, L.-M. Do, T. Zyung, *Adv. Funct. Mater.* **2001**, *11*, 430.

113. C.-T. Chen, *Chem. Mater.* **2004**, *16*, 4389.

114. Y. Yang, R. T. Farley, T. T. Steckler, S.-H. Eom, J. R. Reynolds, K. S. Schanze, J. Xue, *Appl. Phys. Lett.* **2008**, *93*, 163305.

115. Y. Yang, R. T. Farley, T. T. Steckler, S.-H. Eom, J. R. Reynolds, K. S. Schanze, J. Xue, *J. Appl. Phys.* **2009**, *106*, 044509.

116. M. A. Baldo, M. E. Thompson, S. R. Forrest, *Nature*, **2000**, *403*, 750.

117. B. W. D'Andrade, R. J. Holmes, S. R. Forrest, *Adv. Mater.* **2004**, *16*, 624.

118. G. Qian, Z. Zhong, M. Luo, D. Yu, Z. Zhang, Z. Y. Wang, D. Ma, *Adv. Mater.* **2009**, *21*, 111.

119. V. Bulovi, A. Shoustikov, M. A. Baldo, E. Bose, V. G. Kozlov, M. E. Thompson, S. R. Forrest, *Chem. Phys. Lett.* **1998**, *287*, 455.

120. G. Qian, B. Dai, M. Luo, D. Yu, J. Zhan, Z. Zhang, D. Ma, Z. Y. Wang, *Chem. Mater.* **2008**, *20*, 6208.

121. (a) W. P. Gillin, R. J. Curry, *Appl. Phys. Lett.* **1999,** *74*, 798. (b) R. J. Curry, W. P. Gillin, *Appl. Phys. Lett.* **1999**, *75*, 1380. (c) S. W. Magennis, A. J. Ferguson, T. Bryden, T. S. Jones, A. Beeby, I. D. W. Samuel, *Synth. Met.* **2003**, *138*, 463.

122. K. S. Schanze, J. R. Reynolds, J. M. Boncella, B. S. Harrison, T. J. Foley, M. Bouguettaya, T.-S. Kang, *Synth. Met.* **2003**, *137*, 1013.

123. A. O'Riordan, E. O'Connor, S. Moynihan, P. Nockemann, P. Fias, R. Van Deun, D. Cupertino, P. Mackie, G. Redmond, *Thin Solid Films* **2006**, *497*, 299.

124. Y. Sun, C. Borek, K. Hanson, P. I. Djurovich, M. E. Thompson, J. Brooks, J. J. Brown, S. R. Forrest, *Appl. Phys. Lett.* **2007**, *90*, 213503.

125. V. V. Rozhkov, M. Khajehpour, S. A. Vinogradov, *Inorg. Chem.* **2003**, *42*, 4253.

126. J. R. Sommer, R. T. Farley, K. R. Graham, Y. Yang, J. R. Reynolds, J. Xue, K. S. Schanze, *ACS Appl. Mater. Inter.* **2009**, *1*, 274.

127. K. Walzer, B. Maennig, M. Pfeiffer, K. Leo, *Chem. Rev.* **2007**, *107*, 1233.

128. E. L. Williams, J. Li, G. E. Jabbour, *Appl. Phys. Lett.* **2006**, *89*, 083506.

129. T.-C. Lee, J.-Y. Hung, Y. Chi, Y.-M. Cheng, G.-H. Lee, P.-T. Chou, C.-C. Chen, C.-H. Chang, C.-C. Wu, *Adv. Funct. Mater.* **2009**, *19*, 2639.

130. L. J. Kricka, *Anal. Chim. Acta*, **2003**, *500*, 279.

131. (a) M. J. Navas, A. M. Jiménez, *Food Chem.* **1996**, *55*, 7. (b) S. Bezzi, S. Loupassaki, C. Petrakis, P. Kefalas, A. Calokerinos, *Talanta*, **2008**, *77*, 642.

132. (a) D. J. Beale, N. A. Porter, F. A. Roddick, *Talanta*, **2009**, *78*, 342. (b) J. E. Sigsby, F. M. Black, T. A. Bellar, D. L. Klosterman, *Environ. Sci. Technol.* **1973**, *7*, 51.

133. D. Lee, S. Khaja, J. C. Velasquez-Castano, M. Dasari, C. Sun, J. Petros, W. R. Taylor, N. Murthy, *Nature Mater.* **2007**, *6*, 765.

134. (a) Y. Su, H. Chen, Z. Wang, Y. Lv, *Appl. Spectrosc. Rev.* **2007**, *42*, 139. (b) M.-K. So, C. Xu, A. M. Loening, S. S. Gambir, J. Rao, *Nature Biotechnol.* **2006**, *24*, 339. (c) J. V. Frangioni, *Mol. Imaging*, **2009**, *8*, 303.

135. (a) J. A. Prescher, C. H. Contag, *Curr. Opin. Chem. Biol.* **2010**, *14*, 80. (b) A. Roda, M. Guardigli, E. Michelini, M. Mirasoli, *Trends Anal. Chem.* **2009**, *28*, 307.

136. (a) J. M. Baumes, J. J. Gassensmith, J. Giblin, J.-J. Lee, A. G. White, W. J. Culligan, W. M. Leevy, M. Kuno, B. D. Smith, *Nature Chem.* **2010**, *2*, 1025. (b) C.-K. Lim, Y.-D. Lee, J. Na, J. M. Oh, S. Her, K. Kim, K. Choi, S. Kim, I. C. Kwon, *Adv. Funct. Mater.* **2010**, *20*, 2644. (c) B. R. Branchini, D. M. Ablamsky, A. L. Davis, T. L. Southworth, B. Butler, F. Fan, A. Jathoul, M. A. Pule, *Anal. Biochem.* **2010**, *396*, 290. (d) Q. le Masne de Chermont, C. Chané, J. Seguin, F. Pellé, S. Maîtrejean, J.-P. Jolivet, D. Gourier, M. Bessodes, D. Scherman, *Proc. Natl Acad. Sci. USA* **2007**, *104*, 9266. (e) N. Ma, A. F. Marshall, J. Rao, *J. Am. Chem. Soc.* **2010**, *132*, 6884.

137. D. Dini, R. E. Martin, A. B. Holmes, *Adv. Funct. Mater.* **2002**, *12*, 299.

138. (a) H. O. Albrecht, *Z. Physik. Chem.* **1928**, *136*, 321. (b) E. H. White, O. Zafiriou, H. H. Kagi, J. H. M. Hill, *J. Am. Chem. Soc.* **1964**, *86*, 940. (c) E. H. White, M. M. Bursey, *J. Am. Chem. Soc.* **1964**, *86*, 941.

139. (a) K. Gleu, W. Petsch, *Angew. Chem.* **1935**, *48*, 57. (b) R. Maskiewicz, D. Sogah, T. C. Bruice, *J. Am. Chem. Soc.* **1979**, *101*, 5347. (c) R. Maskiewicz, D. Sogah, T. C. Bruice, *J. Am. Chem. Soc.* **1979**, *101*, 5355.

140. (a) B. Radziszewski, *Ber.* **1877**, *10*, 321. (b) E. H. White, M. J. C. Harding, *J. Am. Chem. Soc.* **1964**, *86*, 5686.

141. (a) M. M. Rauhut, D. Sheehan, R. A. Clarke, B. G. Roberts, A. M. Semsel, *J. Org. Chem.* **1965**, *30*, 3587. (b) F. McCapra, D. G. Richardson, *Tetrahedron Lett.* **1964**, *43*, 3167.

142. (a) E. A. Chandross, *Tetrahedron Lett.* **1963**, *4*, 761. (b) M. M. Rauhut, B. G. Roberts, A. M. Semsel, *J. Am. Chem. Soc.* **1966**, *88*, 3604.

143. A. Natrajan, D. Sharpe, J. Costello, Q. Jiang, *Anal. Biochem.* **2010**, *406*, 204.

144. A. Natrajan, Q. Jiang, D. Sharpe, S.-J. Law, US Patent 6,355,803 B1 (2002).

145. J. M. Baumes, J. J. Gassensmith, J. Giblin, J.-J. Lee, A. G. White, W. J. Culligan, W. M. Leevy, M. Kuno, B. D. Smith, *Nature Chem.* **2010**, *2*, 1025.

146. J. J. Gassensmith, J. M. Baumes, B. D. Smith, *Chem. Commun.* **2009**, 6329.
147. (a) M. M. Rauhut, B. G. Roberts, D. R. Maulding, W. Bergmark, R. Coleman, *J. Org. Chem.* **1975**, *40*, 330. (b) K. Kimoto, R. Gohda, K. Murayarna, T. Santa, T. Fukushima, H. Homma, K. Imai, *Biomed. Chromatogr.* **1996**, *10*, 189.
148. G. Qian, J. P. Gao, Z. Y. Wang, *Chem. Commun.* **2012**, *48*, 6426.

Index